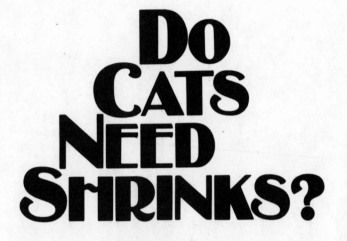

Do Cats Need Shrinks?

Do Cats Need Shrinks?

CAT BEHAVIOR EXPLAINED

PETER NEVILLE

CASTLE BOOKS

Library of Congress Cataloging-in-Publication Data

Neville, Peter (Peter M.)
    Do cats need shrinks? / Peter Neville.
       p.   cm.
    "First published in 1990 by Sidgwick & Jackson Limited,
London"—T.p. verso.
    ISBN 0-8092-4051-3 : $13.95
    1. Cats—Behavior.   2. Cats—Psychology.   I. Title.
SF446.5.N48   1991
636.8—dc20   91-6372
                                 CIP

The publisher would like to thank the following doctors from Carlson
Animal Hospital in Oak Park, Illinois, for their contributions: Richard
A. Carlson, D.V.M., Michael J. Forret, D.V.M., Andréa D. Tasi, V.M.D.,
Lucia Lemmer, D.V.M., and Robert A. Badertscher, D.V.M., Ph.D.

This edition copyright © 1994 by Castle Books
a division of Book Sales, Inc.
Raritan Center
114 Northfield Avenue
Edison, NJ 08818

Printed in the United States of America

ISBN: 0-7858-0158-8

To Claire. It's time I popped the question.

Contents

Preface

It was with great delight that I learned that *Do Cats Need Shrinks?* would be published in the United States. It was already selling well in the United Kingdom and in other English-speaking markets such as Australia, but to publish this type of book in America was especially exciting for me. You see, the practice of behavior therapy for pets was founded in the United States some twenty-five years ago and has since spread across northern Europe and as far as Australia.

Originally, of course, such therapy for pets was viewed as outrageous. But the bond we all share with our pets has proved to be so strong that now, just a few years later, taking your cat or dog to the shrink has become more than just acceptable; it's become a must for those who really care about their pets and who are having a few problems with them. Many European therapists, myself included, have built on the original ideas of American practitioners, and we now communicate frequently with each other by post and at international meetings; thus we all develop our abilities to do the job better. It is particularly pleasing now for me to present both to my American colleagues and to all American cat owners exactly what I have done in developing the art and science of feline behavior therapy.

Cats, of course, are much the same the world over, though their relationship with us and our expectations of them can vary from culture to culture and from country to country. They all have their own personalities, idiosyncrasies, and occasional problems. This book is designed to help their owners understand their relationships with their cats a little better from their cats' point of view and to offer frontline advice to help them when behavior problems arise. Of course, many cats will still need a more personal approach to treatment, and it is comforting to know that there are many professional behavior therapists across the United States who can offer that help. They are, I hope, following the pattern established by my practice and others in the United Kingdom of seeing cases only on referral from a veterinarian. After all, if you were having a few psychological problems as a result of being medically ill, you'd rather be seen by a physician than a shrink, wouldn't you? Well . . . so would your cat!

To all owners and admirers of cats who read this book, I hope you find help and interest in its pages and are reassured that cats the world over may be sharing the same problems and, with a little help, overcoming those problems. Be assured also that here in England, we feel the same about that most magnificent and affectionate of pets, the cat, and that I and my colleagues are continuing the work begun by your countrymen.

Peter Neville
Wiltshire, England
January 1991

Acknowledgments

With many thanks to Claire for all her patience, love, and skill and to Bullet, Scribble, Wandsworth, Nimrod, and Thumper, the cats in my life so far. Thanks too to all those veterinary surgeons who refer cases to me, especially to Ian Hughes and partners, Keith Butt, Geoff Little and partners, and Professor Keen and Tim Gruffydd-Jones and colleagues, who provide such excellent facilities and expertise to support this pet shrink in his work. Thanks also to my long-suffering parents, who have to tell their friends what their son does for a living, and finally to cats and their devotees everywhere for all their inspiration.

Introduction

*If man could be crossed with the cat it would improve man,
but it would deteriorate the cat.*

Mark Twain

"Mr. Neville? I'm sorry to phone you so early on a Sunday,
but it's my cat . . . no, it's my husband really. He says the
cat's got to go. It's done it once too often, and he says it's
either the cat or him . . . and you're my last resort."

So begins another Sunday at 7:30 A.M. Gently, I edge
my own usually hyperactive and hypervocal, but now thankfully asleep, Siamese cat Scribble and her ex-feral friend,
Bullet, to one side and crawl out of bed into my office to
grab the other phone.

"Now, Mrs. . . . ."

"Fisher."

"Let's work through this logically so that I can get a
better picture. First, what sort of cat do you have, what sex is
it, and how old is it?"

"George is a girl. We thought she was a boy, but she isn't.
She's a one-and-a-half-year-old moggy. We've one other cat,

and they seem to get along okay, but she just won't stop doing it!"

"What is she doing, Mrs. Fisher?"

"She pees everywhere. But it's not *peeing* peeing because she's standing up. She also does puddles behind the sofa, and my husband says he's had enough."

"How long has George been doing this?"

"About three months, ever since the cat next door came in and had a fight with her in the kitchen. Oh, please say you can help; I couldn't bear to have her . . . well, you know, at the vet's. . . ."

It's a typical case. A normal, much-loved cat has suffered the trauma of invasion by a local rival right in the safe core of her territory. The incident has long passed, but the consequences on poor George's behavior continue. It is a relatively common case of nervous urination in a safe place indoors rather than going outside, where she is too vulnerable when squatting. Spraying the house is a desperate attempt to keep a hold on those resources she cannot leave—food, love, and shelter. The much-loved cat has quickly become unloved by one member of the family, and relations between the owners are clearly strained. I may be the last resort. The fate of the cat and perhaps even the marriage may hinge on my successful treatment of the problem.

First, I ask Mrs. Fisher to talk to her vet to make sure that there are no medical reasons behind George's behavior and to ask him or her to refer the case to me. Then I'm off to town to make the necessary house call at the Fishers' and take a closer look at the problem. Two hours later and we've worked out why George is behaving the way she is and compiled a program of treatment involving modification of the home environment, controlled exposure to sensitive areas, and restructuring of the relationship between the owners, especially Mr. Fisher, and George. Within a couple of weeks at most the problem should be resolved.

As I was leaving the Fishers' to enjoy the rest of my Sunday, my client casually remarked, "George has been munching her way through the back of the bedroom chair since she was a kitten. Is there anything you can do about it?"

Not content with being a sprayer and a nervous toileter, George is about to form part of my continuing research into the phenomenon of fabric eating by cats. Now that *is* worth a bit more of my Sunday.

On rare quiet days I often wonder how it came about that I should be a "pet shrink." Having passed through the usual early ambitions of footballer and explorer and then developing my lifelong interest in animals by studying biology at university, I still could never have predicted that I would become a professional consultant in pet behavior. Nor could I have known that my Sunday mornings would be spent treating spraying, toileting, fabric-eating cats. That it should happen to me is perhaps at least understandable, given my attachment to animals, but that our society should have a place for a consultant in feline behavior is undoubtedly surprising to many, not least to those of my father's generation. Thirty years ago a cat was an animal that scarcely warranted feeding, let alone veterinary care or treatment from a cat psychologist.

So times must have changed, and our relationship with our cats must have altered, for me to do what I do. This book is about the types of problems that I treat and the people and cats that I meet. Above all this is a book for those who, like me, are incurably fascinated by cats; it also provides an insight into the more unusual aspects of their behavior and their relationship with us.

⟍ 1 ⟋

Cats, Dogs, and People

Cats have rarely been of much use. Aside from the often overstated mouse- and rat-catching role, cats have not followed the pathway of domestication based on training to task followed by the horse or the dog, nor the pathway of providing us with food (with certain Eastern countries excepted) as have farm animals such as the cow, pig, and chicken. Nonetheless cats have always been with us—worshiped in ancient Egypt yet sacrificed as accomplices to witchcraft in Europe in the Middle Ages. Rejected and doted on, the cat is now even more popular as a pet in the United States than man's traditional best friend, the dog. The same is set to happen in the United Kingdom and probably elsewhere in northern Europe by the mid-1990s.

Despite this popularity we have achieved little success in altering the basic design of the cat compared with the way we have manipulated dog genes to produce such breeds as the Chihuahua, Great Dane, Mexican hairless, and dreadlocked Hungarian puli. Instead it is human beings who have changed: physically we are on average larger than at any other point in history, and socially we are changing even

5

faster. The cat, a perfectly evolved top predator, is completely suited to come with us into our modern, self-created
technological niches. People like cats because they are clean
in small modern apartments. They are self-determining and
self-amusing and loved especially because they are affectionate to us in an increasingly formal and businesslike world
that affords little opportunity for emotional expression. Children are growing up ever earlier these days and starting to
reject the nurturing, protective behavior of their parents far
sooner than they did twenty years ago. In contrast, the cat is
there to love and will always—well, nearly always—respond
to our desire to initiate affection.

In a world where security means a written contract and
emotional expression is often seen as weakness, the friendly
cat is one of our only real-life outlets for uninhibited expression. How reassuring that on our return home the cat will be
pleased to see us, purr, and rub around our legs in an
unconditional, uninhibited display of affection. And how
reassuring that we can feel free to respond to that display
without our fellow man belittling us or perceiving our expression of emotion as an opportunity to take advantage while
our defenses are down. Being affectionate with your cat or
dog is socially acceptable. And nowadays people need that as
they never have before.

Why Not Dogs?

*To his dog, every man is Napoléon, hence the constant pop-
ularity of dogs.*

Aldous Huxley

Our association with the dog goes back approximately
10,000 years. The relationship was based primarily on the
dog's willingness to assimilate into our social group in return for the benefits of food, shelter, and protection. We
behaved as a dominant pack leader toward him, and he was
prepared to adopt the subordinate role at the bottom of our

social hierarchy. The benefit to us was that the dog would, and of course still does, share in the role of protecting our common interests: at first fending off the wolves from our livestock and then guarding our homes and shared territory. Even today a dog's bark is seen by many insurance companies as such a deterrent to robbery that they offer cheaper home insurance policies to dog owners.

Dogs were often kept for their speed and skill in hunting, thus improving our survival prospects. Nowadays their physical skills of scent following are cleverly applied to drug and explosives detection work, and their ability to be trained is invaluable to the blind, to the disabled, and to the police in apprehending the ever-increasing numbers of villains. But while it is true that the number of dogs being kept in the United Kingdom is rising at a rate of about 10 percent per year, fewer and fewer of us can afford the luxury of a dog simply as a companion, as man's best friend. As in earlier times, the dog must once again do something to earn its keep and justify our patronage. The popularity of the larger, tougher guarding breeds to protect us and our property perhaps indicates that the modern role of dogs in our cities is increasingly returning to the utilitarian as it was all those thousands of years ago. It's just that we—not the wolves—are now our own worst enemies.

Protection in the form of a hefty rottweiler or barking German shepherd may be necessary for survival in some areas of our cities not sufficiently wealthy to afford intercoms at the door or full-time surveillance systems. But we should already feel uneasy at encouraging dogs to be aggressive and worry about how controlled that aggression is. We also worry about our environment and the use of parks and open spaces as dog toilets. When this extends to pavements, the outcry is even louder—there is nothing worse than stepping in it.

Add to that fears of enteric worm and other diseases transmissible from dogs to people via their feces and we have a pet under pressure in our crowded cities from the "green" point of view. However good they are at protecting us, dogs are getting more and more unpopular with the non-dog-owning public, and calls for tighter regulation of ownership, licensing, and physical control are heard frequently in the press and from politicians.

Cats too can infect us. They can carry enteric worms and deposit their feces outdoors in soil or, less desirably, in children's sandbox play areas, where worm larvae may be inadvertently ingested. While the infection of humans by the roundworm *Toxocara cati* is as yet unproven, compared with the established infection with *Toxocara canis*, the risk of exposure to a protozoan parasite causing toxoplasmosis from a feline is higher. Over 50 percent of the population will be infected at some time, yet few of us ever develop symptoms. The real risks are to women who become infected when pregnant, causing miscarriage or damage to the fetus. Ensuring that cats are wormed regularly and that extra care is taken when cleaning litter trays or gardening will minimize the chance of infection. (The main risk is from meat that is not cooked thoroughly—so keep off the extra-rare when pregnant.)

In cities it is the wealthier, younger members of British and American society, the yuppies and the dinks (double income no kids), who lead the way in the keeping of cats. They can often afford those noncanine security systems for their property and so need a pet for more emotional reasons. Cats are kept by the busy younger age group but also in owner-occupied houses where the head of the household, male or female, is employed full-time. Thankfully not all pet owners are alike or easily categorizable and not all well-off people live in cities. Many people decide to keep no pets at all or, like this pet psychologist, have both cats and dogs (just in case you thought I was being antidog in any way).

The well-respected Yale-based human/animal bond philosopher Stephen Kellert proposes that all pet keeping falls into four categories. There are those in our society whom he describes as "negativistic," who dislike or are afraid of animals and actively avoid them. Many of this group have pets thrust upon them later in life and find they have been missing out on something rather special. However, as few "negativists" will be reading this, I'll pass on to the next category, which is "utilitarian." Utilitarians' main concern for animals centers on their practical or material value. Older members of our society predominate in this class, having been brought up in harsher times when every mouth to be fed had to contribute to the family in some way.

While many would-be utilitarians in the Western world have no need to keep pets now, younger people in the 18-to-35 age bracket can enjoy being "humanistic" or even "moralistic" because present-day wealth allows them the indulgence. Humanists are interested in and have a strong affection for individual animals, and most pet owners nowadays fall into this category. Moralists enjoy the further luxury of being concerned with the rights of animals and are opposed to all forms of cruelty from a moral and often campaigning stance. The moralist attitude is almost incompatible with the utilitarian ideal and, of course, unthinkable in the Third World or even among the worst-off in the Western world.

It is perhaps a paradox that many elderly utilitarians are now benefiting from programs of pet-facilitated therapy designed to improve the lives of our institutionalized elderly, physically disabled, and mentally ill. Cats and dogs are increasingly being taken into hospitals and even penal institutions to help rehabilitate or simply provide some undiscriminating love and affection. We all know of the relaxation benefits in lowering heart rate and blood pressure derived from even the simple action of stroking a cat or dog, and perhaps we have designed a society where we have come to depend on that facility.

For modern man, life is getting too unpredictable to keep a dog. Dogs have to be walked, exercised, and fed at set times. They can't be left alone at home for too long or accidents happen. Nowadays both partners in a relationship usually have to work to pay for mortgages and so are out all day. Late or variable working hours and plenty of places to go for fun afterward would just mean a miserable life for the dog. Yet after a hard day and evening at the office, an after-hours business meeting in the wine bar, a trip to the theater, and a late-night drive home, making the obligatory call to relations and abandoned friends on the car phone, modern man needs his pet. He has to stay human somehow! Increasingly the pet of choice will be the cat.

The director of the Companion Animal Research Group at Cambridge, Dr. James Serpell, argues that, in the past few decades, not only have we shifted our relationship with animals—except those large guard dogs—away from the utilitarian approach of many of our parents, but also our increasing humanism has allowed us to accept and develop strong emotional attachments to our pets that would previously have been regarded as unhealthy or odd. An inevitable consequence of this close relationship and the lack of human company in big cities (you're never so alone as when you're alone in a city, surrounded by people) is the viewing of our pets as people, the process of (deep breath) anthropomorphization.

While small cage pets such as hamsters are also increasing in popularity, they are simply not highly developed enough for the type of emotional involvement we now seek. And you can't really have an exchange of emotions with a tankful of fish, no matter how aesthetic and calming to watch (I hope your dentist has a tank in his or her waiting room), so again we can think cat.

Cats do well as apartment creatures despite being so highly adapted as territorial predators. That's why there are more than 55 million in the United States, where over 80 percent of people live in cities, and more than 6.7 million in the United Kingdom, where more and more of us live in cities or suburban areas. Easy to keep, adaptable, and happy to enjoy what we offer without depending on us, the cat was never so perfectly suited to us. Cat haters, of which there are still many (few are ambivalent toward cats—you either love 'em or hate 'em), often despise the cat's very independence and desire to be fed, sheltered, and loved on its own terms while remaining aloof. They may prefer the slavish devotion of the dog. In biological terms the cat's relationship with us could be described as parasitic, compared with the symbiosis of our relations with the dog, but that would be to place no importance in economic or survival terms on keeping cats, when we now need them more than ever before for our emotional well-being.

Cats have rarely been of much use, yet they have stuck with us throughout thousands of years. And as that relationship evolves, new demands and responses start to emerge. The resultant sanity of cat and owner is what this book is all about.

The Relationship

The earliest known forerunners of the present-day domestic cat appeared on earth about 13 million years ago, with the larger forms such as lions and tigers not arriving until about 10 million years later. In between times the cat was so successful at the top of the food chain that it spread around the world to all parts except Australasia, Antarctica, Madagascar, and the smaller islands.

The most likely ancestor of our modern pet, *Felis catus*, is the reddish to gray-brown African wildcat, *Felis lybica*, which is slightly larger and has longer legs, a leaner body, and a long, thin tail. This species tames readily, and the earliest records of man's association with any cat date from 2600 B.C. in ancient Egypt, which was part of the African wildcat's territory. Evidence of feline domestication can be found from 1600 B.C. onward in Egypt, where there are many tomb paintings depicting the cat's role in home and family life. By this time the Egyptians had already been domesticating other species for more than 1,000 years, so perhaps the late arrival of the cat in Egyptian homes is due to social changes. The more advanced a civilization be-

13

comes, the better it cares for its weak, and therefore the
more popular the practice of keeping pets becomes. It is also
possible that the cat underwent a genetic change that made
it more docile and tolerant of man, enabling it to come into
his house from the outside. Early handling of young born
indoors facilitated taming, and thereafter followed a deliberate selection of friendlier, responsive individuals.

This is largely supposition, of course, deduced from what
little archaeological evidence is available, but while the African wildcat was able to evolve into a domesticated species,
by contrast the pure tabby European wildcat, *Felis sylvestris*,
has never been tamed. As the African wildcat spread or was
transported as a pet by man, it interbred with its European
cousin to produce the original tabby markings of what we
now know as the domestic cat.

Domestication itself implies a genetic disposition for
docility that man is able to exploit and then deliberately
select in breeding to make his animal companions nonag-
gressive, less fearful, and easier to handle. Clearly the modern farm species were originally selected for production, but
docility was also essential if they were to be enclosed and
bred efficiently and safely. Dogs too were selected for docility and function, and it was only after thousands of years of
producing tame dogs that we really started to manipulate
the dog's genetic complement for appearance and to produce the 450-plus breeds we know today. Most dogs still
fulfill basic guarding roles for us, but a few are kept for
specific tasks such as hunting or herding sheep. They are
with us as companions and in many ways have been so
altered by us in the process of domestication as to have little
chance of surviving without us. There are plenty of feral (wild
or domestic gone wild) dogs throughout the world, from New
York to Liverpool to the high regions of Italy, but these do
not survive long unless able to group with others and re-form
a successful hunting/scavenging unit as their ancestors did.
Certainly the most successful feral dogs are the medium-size
to large mongrels with short- to medium-length coats, as
dogs of the artificially maintained pedigree strains soon

interbreed if they survive at all away from our direct care.

Domestication is therefore something of a compromise for many species, but less so for the cat. The loose association with man from earliest times, when the cat was encouraged as a rodent controller around our settlements but not assimilated into the human family as part of the "pack," has ensured that its nature has not been fully compromised by our efforts. Since domestication, cats have been allowed into our homes, but they remain free to come and go as they please, so ensuring maintenance of their self-determining behavior patterns. Of course we have selected in a general way for tameness, but having evolved as a solo predator, the cat has no need of us for actual survival and hence no need to accept a subordinate position in our family for the benefits of food or shelter. As an opportunist the cat takes advantage of the food we offer and appreciates the shelter, using our house as a lair, but always retains the ability to fend for itself. The process of domestication never really touched that successful evolution as a hunter, and all cats (perhaps with the exception of the squash-faced Persians) remain proficient and active predators given the opportunity or necessity. Even the best-fed cats still bring home birds and mice.

There are huge numbers of feral cats surviving and breeding to maintain populations, so they do not rely on immigration from stray or abandoned pets. Ferals are found in all corners of the world from the Galapagos Islands (where they have become a serious threat to the unique fauna) to the hotels of the Mediterranean and Indian Ocean coasts, to the famous cats of the Colosseum in Rome, the streets of London, and all cities in East and West, as well as on many farms. The domestic species reverted to a wild lifestyle seems to do far better than any of the original wild species, such as the Scottish wildcat, which are mostly in decline due to persecution by man and loss of their habitat. This is probably due to their willingness to stay close to man for ease of finding food. Some feral cats, such as those populations arising from cats abandoned on uninhabited islands by passing ships, do survive solely by hunting; others exist

largely by scavenging from garbage bins or from handouts
provided by kindhearted people.

The domestic species of cat is therefore not totally dependent on us. Instead it survives extremely well on a range
of feeding strategies when nearly all other species we have
taken on are now too altered to survive without us. The
reason they stay is because we make life easy and all that
domestication has done is to select those cats that are better
at learning how to get the best from us by displays of affection and relaxed companionship. But this is to paint too
simple a picture of the cat's side of that special relationship
with us. Uncompromised perhaps, independent certainly,
but the huge bond we are able to establish with such a highly
developed creature is not just a case of détente or mutual
acceptance. Instead we must look to the developmental
biology of the cat to discover where we fit into the cat's
perception and why we are more than just tolerant cohab-
itants.

For the dog we are the pack leader, a Napoléon who
makes decisions, orders life, protects the den and other pack
mates, and allows access to food. This relationship is incon-
ceivable with a nonpack animal. To a cat we are its mother,
long after it would have become independent and, in fact, for
the whole of its life. The only time in a cat's life when it would
ordinarily be dependent on other members of its species is
as a young kitten, when it needs its mother first for milk and
then to bring home more solid food when it is weaned. By
this stage the kitten is already showing marked independence through exploratory behavior and is having progressively less contact with its littermates. At only twelve weeks
after birth it is able to fend for itself and is a fairly proficient
hunter.

By contrast the twelve-week-old dog is very much a
juvenile and still learning how it should behave within the
rank-order system of the dog pack. It must learn to wait for
its share of the spoils of hunting and that older, high-ranking
dogs take precedence in selecting favored shelter and walkways and have the right of way. Taking a puppy from its

mother after weaning at six to twelve weeks and placing it in a human pack is simply to divert the usual assimilation into the dog group. It is noticeable that dogs placed in a human home early after weaning (at about six weeks), when handled frequently and, after vaccination, exposed to as wide a variety of people, other dogs, and environments as possible, are far less likely to present behavior problems when growing up or as adults. Lack of sociability with other dogs or people due to nervousness or aggression is largely avoidable given the right sort of early exposure. Keep a puppy isolated during that crucial period of six to eighteen weeks old and it will inevitably be incompetent later.

The same principles apply generally to the socialization of kittens. Frequent early handling and placing in a human home soon after weaning at six to eight weeks is most likely to produce tractable, playful, and friendly young cats, even if they are still destined to be territorial and not very sociable with other local cats. After all, none of the small wildcat species is particularly sociable in the wild. Yet if we take feral kittens born in the wild and unhandled before about eight weeks of age, we can quickly tame them and make them into pets that are completely indistinguishable from home-raised kittens. However, take a six-to-eight-week-old Scottish wildcat, and it is beyond taming; instead it follows the rejection patterns and developmental independence pathway we have all probably experienced in trying to raise young wild rabbits, squirrels, or other mammals. While they are young and dependent on us, they are friendly, accepting us as a mother substitute. Once nutritionally independent and entering adolescence, they reject us and move on to establish their own territory away from the "maternal" home.

Even more interesting than the prospect of taming feral kittens is the fact that many adult feral cats that have never been handled or had any contact with man can also be tamed into perfectly friendly and responsive pets. Achieving it takes a little longer than with feral kittens and may require a concerted effort and the use of intensive socializing techniques, but it is being done by cat rescue groups all over the country. However, the same efforts with a true wildcat barely achieve recognition behavior between man and cat.

The feral cat is able to live in a wide variety of social systems, depending on the availability of food. Where food is scarce, feral cats are likely to be solitary and intolerant of other cats, but as food supplies increase, their capacity to be sociable rises. It is probably not worth risking injury in a fight when there is enough for everyone, so large numbers of cats may be found together, all apparently tolerant of each other. While many groups comprise related animals in an extended family arrangement, others are made up of unrelated feral cats that sometimes develop a complicated system of social organization. They rarely, if ever, hunt together or share prey, yet they will eat together peacefully, and one mother may suckle another's kittens and share in their protection. It seems that domestic cat society is highly variable and extremely elastic—another feature that has surely contributed to its ability to assimilate into so many different human lifestyles. Perhaps it was sheer chance that the African wildcat possessed or developed the genes that allowed it to be tamed by the Egyptians, and perhaps we owe the success of the pet cat today to that chance, for without it all cats would possibly still be shy, distant creatures and declining in numbers like most of the small wildcat species.

So our modern domestic cat is equipped to accept us and live with us. But this can be achieved only because we fulfill the role of its mother and mimic the only time in its life when the cat is truly a social animal. We not only provide food when the young adult cat would be expected by its mother to find its own but also allow the cat to demand physical attention, warmth, and affection from us as only its mother could provide. We offer the same security in our laps as the kitten knew when suckling or lying next to its mother. Often the cat responds with purring and affectionate rubbing, and also by kneading our laps as it used to knead its mother's nipples to stimulate milk flow. Some particularly affectionate or closely bonded individuals even salivate and dribble at these times in anticipation of a milk feed. Dribbling perhaps excepted, we love this infantile display of affection and cannot resist the urge to stroke our cat and be affectionate in return while sitting with it on our lap or cradled as a baby in our arms. Rarely is the relationship one of "child substitution" for us, yet relaxed, warm, and protecting, we all play the role of a cat's mother throughout its life and, as a result, relax and feel better ourselves.

At other times we play with the cat in simulating hunting by offering moving targets of Ping-Pong balls and string to chase as its mother would have enticed it with her tail to teach it to stalk, chase, pounce, and master other predatory skills. As the kitten was weaned, its mother would have brought back half-killed small prey for it to learn to catch and dispatch and thus further learn the hunting game. Instead we provide full meals, already dispatched, that simply need eating, so enhancing the providing mother image. Small wonder cats are so affectionate at mealtimes. Lastly we provide a safe lair—our house—where the cat is protected by four walls (and a whole family of mothers) from enemies and competition with other cats. This is also the place where those meals and affection are provided, so it is always worth returning to, unless of course there is better service down the road.

Our cats need us, and we encourage the dependence constantly, but the maternal relationship is as much a function of cats' behavior toward us as ours is to them, and even when totally indulged as true "child substitutes," they retain the ability to grow up quickly and survive well without us. We build up enormous emotional attachments to our cats— there is little option when playing a maternal role—but more than this, we are starting to need to care for something in our modern and increasingly emotionally vacuous society, and the cat is the perfect receiver for the release of our maternal and loving emotions.

Cats and Culture

Our bond with the cat has developed from loose attachment for the benefit of rodent control, through tolerance of a noncompetitive self-reliant creature, to dependence, but throughout the ages the cat has also been admired simply for its grace of movement and beauty of form. In the closer modern relationship that admiration has become more pronounced. Poets and artists have exploited feline attributes since Egyptian times, and the languid shape of the cat was a major theme in the Art Nouveau of the 1920s in the West.

Nowadays we are all at least armchair naturalists. The modern media of television, cinema, and highly advanced photographic techniques have brought the diversity and complexity of nature into our homes and enthralled us all. As a result we are more educated and no longer regard the predators at the top of the food chains—such as lions, pike, and eagles—as ruthless, harmful killers. We can now understand their role in keeping nature fit by feeding on the weak, sick, and old and as a result can view the predatory side of our pet cat if not with pleasure, then at least with understanding. We still marvel at its graceful ease of movement and sharp reflexes, but we also understand the evolutionary benefit and functional adaptation of the various features of the hunter. Highly developed senses and efficient weapons of claws and teeth are all part of the divine plan. This opportunity to have the tooth and claw of nature contrasting with its absolute beauty and benign character as it lies by our fireplace in the city is yet one more reason for the increasing popularity of cats in human society. Urban tigers one minute, babes in arms the next—a perfect contrast.

Do Cats Need Shrinks?

Cats represent extremely good value in terms of the pleasure we get from them, as each cat has two quite distinct characters.

Indoors the cat is affectionate, placid except for odd bursts of activity, and manages its life around us without getting in the way. Outdoors it snaps back into its wild alter ego and becomes the hunter once again. While we may disapprove of its successes, we still stand and stare enthralled at its stalking skills and its tenacity in defending its patch of garden territory. Home and hunting ground defended, birds put to flight and mice to ground, the cat pops back home and reverts to its kittenhood at the drop of a hat.

The vast majority of cats live out their nine lives in this very attractive style, which couldn't have been better designed to suit their every whim. A few highly prized pedigree cats, those at risk near busy roads, and increasing numbers in our cities are denied the opportunity to express their wild component and are kept indoors for all their lives. While this may seem a pity in that the owners never get to see the uncompromised predator in their cat, the animal itself rarely seems bothered. Provided the cat is taken in as a kitten and never allowed to experience the outdoor life, it can adapt perfectly well to apartment life. A case of "what you've never had you don't miss" applies to most, although it would be unthinkable to try to confine a cat to the indoor life if it was used to patrolling a territory outdoors. Frustrations would understandably develop and alternative, and perhaps unwanted, behaviors arise, such as pacing, late-night crying to be let out, and overreactions to minor events, such as chasing and biting passing legs.

At last a mention of the possibility that our cat's behavior could sometimes be less than perfect. The solution may be obvious in this case—either take the risk and give your cat back its established freedom or find it a new home so that it can start to enjoy life again. But for other problems with less obvious causes, treatment may prove difficult. For the cat that soils your carpet, sprays your kitchen counters, or runs and hides as soon as the doorbell rings, the quick solution of rejection by finding the cat a new home or asking the vet to end it all may be the only prospect after you've tried all the best advice of friends, cat breeders, and the local vet.

But most of us are far too involved in that maternal relationship for such callous expediency with a creature that, at its worst, is perfectly well behaved for 99 percent of the time and one on which we have come to depend for more than just companionship. Small wonder that most owners of cats with behavior problems put up with them.

Owners often blame themselves for their cat's behavior problems and feel that their cat is simply reacting to a lack of love or their failure to provide the right type of home environment. They may worry that they are inadequate owners rather than let the blame fall on the "innocent victim." As a result the emotional compromise of loving a "rotter" who continually abuses all that love and care can place enormous strain on owners and their families and, ultimately, far outweigh the joys of owning the cat in the first place. Owners of house-soiling or indoor-spraying cats invariably reduce their human social contacts because they are embarrassed to let people see their unclean homes, and so their dependence on the cat for company rises still further and any desire to dispose of it wanes. The circle becomes vicious and the owners more and more frustrated, until eventually family relations may start to suffer and the owners are forced to take some sort of action.

But just what constitutes a problem is highly variable. For some a single mistake on the carpet when their cat is "caught short" is sufficiently damaging to their enjoyment of keeping it to cause rejection. Most put up with the same event as something to be expected from time to time and accept that what with fur and footprints owning a cat or a dog inevitably means accepting a lower standard of home hygiene. Many others put up with far worse problems without turning a hair, almost regarding their cat's unhygienic lack of toilet training as normal. The definition of a problem is a highly personal matter based on owners' individual tolerance levels and their relationship with their pets.

A behavior problem is often acknowledged only when the pains of coping with a cat start to exceed the joy it brings. If the owner of a cat finds any aspect of its behavior or their relationship problematic, it ill behooves anyone to tell the owner that he or she is worrying unnecessarily. But just where owners can turn for help to tackle any problem is rather limited. First they may be unwilling to reveal that they have a problem with their pet because of the guilt factor and the necessity of letting it be known that all is not well and that, by implication, their pet is less than happy with them. They may be particularly unwilling to disclose house-soiling or indoor-spraying problems because of the reflection on the standards of cleanliness in their own home. But if the problem simply isn't responding to the owner's efforts to resolve it, they must come out and admit that they are sufficiently attached to their cat to suffer in putting up with its problems.

Owners usually turn for advice first to the relative safety of friends or relations who also have cats and then to their cat's breeder or a well-known local cat fanatic. The vast

majority of problems are resolved by such help or, if not, at
the next port of call—the vet's. Experience often tells, particularly with house-training problems in young cats, but there
is a small minority of problems that are either so rare as not
to have been experienced before by even breeders or vets or
are so complicated that there is no easy or straightforward
solution. Then there are other persistent problems, including
many indoor sprayers, which either don't respond to treatment or do so only temporarily, and relationship problems
where the cat simply isn't responding to the owner's desires
or expectations, perhaps because of nervousness or low-
threshold aggression. Tackling these types of problem, be
they of feline behavior, relationship incongruities, or, as it is
often the case, a combination of both, takes time. Time to
look at all the factors that may be influencing the cat's
behavior, from the structure of its home environment to its
relations with each member of the family and its friends or
rivals from down the street. Time to look closely at the cat's
development and early upbringing in the true psychologist's
style of going back to childhood and looking at its general
character and disposition aside from the problem. No two
cats, no two owners, and no two problems are exactly alike,
so a very broad approach is required to identify likely causes
of each problem and to develop a personal treatment plan
with each owner. It takes time, a good background in feline
and general animal behavior, and, most of all, a caring
approach in dealing with a very important and sensitive
aspect of people's lives. Lastly you need to be able to moti-
vate owners to tackle the problem when they may be nearing
the end of their own patience.

It sounds like a blueprint for a consultant in feline
behavior with just a touch of human psychology and counseling skills, and it is. Do cats need shrinks? Perhaps they
themselves don't, and neither do most of their owners. What
they do sometimes need is a feline behaviorist with the time
and patience to understand why the cat is doing what it's
doing and when and how to modify it within the framework
of the cat's relationship with its owners and the geography of

its home territory. The relationship we have with our cats can also need adjusting from time to time, and owners can be given a little advice on how to restructure that relationship to bring about changes for the better in the cat and alleviate certain types of problem behavior. The cat's behavior can need only a little improvement, or sometimes a lot, for owner and cat to appreciate life and each other to the full once again or sometimes even for the first time.

And the more we enjoy keeping cats and the more we find emotional release in them, the more concerned we will become when things do not run smoothly. As in so many other areas, we increasingly look for expert help from a detached professional viewpoint to help us iron out the difficulties when our own efforts fail to resolve those problems. The days of punishing or rejecting a cat for the crime of having a behavior problem are over because help is now available, and given the right approach those problems can usually be resolved. If the strength of relationship between man and cat wasn't sufficiently strong even ten years ago to warrant a consultant in feline behavior, it is fast becoming so now.

In Practice

My practice runs strictly on referral from veterinary surgeons, and I work almost exclusively alongside veterinary surgeons at their premises. Obviously a cat's behavior, like ours, will change when it is ill, so it is important that the cat first be examined by a veterinary surgeon to eliminate this possibility before I see the case. Other cases, such as cats that mutilate themselves, are best seen by a combination of vet and behaviorist as the behavior may arise from influences in both camps and treatment involves a combined approach. But ordinarily I see cases in private with the owners and their cats for approximately one to two hours at two veterinary hospitals in the northwest of England (where I see dog behavior referrals as well), at the University of

Bristol Veterinary School, and at a veterinary practice in southwest London. I also make a few house calls for owners who have difficulty traveling to those centers. During each consultation I record details about the nature of the problem and the cat's and owners' lifestyles. Having discussed the problem, I then try to work out a program of treatment for the owners to carry out at home. After the consultation I forward a written report to them, and a copy is sent to the referring veterinary surgeon together with any suggestions for drug treatment such as the prescription of sedative or hormone treatments that I feel would assist treatment. The owners stay in contact with me by phone, and we reassess treatment after two to three weeks. Only rarely will I need to see a case twice.

I always count myself as very lucky because not only do I get to work with my favorite animals professionally but I also get to meet the nicest people. Pet owners are more sociable than non-pet owners, and whatever their pet's problems, cat owners especially are always pleasant, sensitive, and interesting. More importantly, they usually have a splendid sense of humor, and this is reflected in the following chapters, where we will be looking broadly at the types of cases I treat and referring to past cases, typical and atypical. Only the names have been changed to preserve client confidentiality and to protect the innocent.

Problems, Problems:
Nervousness, Phobias, Insecurity

Dear Mr. Neville,

I have two female spayed cats, not related, one four and one five years old. The four-year-old, Sammi, is an extrovert, totally confident and very affectionate with everyone, including the dog. Shelley, by contrast, is timid, runs and hides from every noise, and creeps around in fear all the time. They both go out, but Shelley never goes far and runs indoors at the slightest disturbance. In fact she's really at ease only on my lap. Should I worry? Is there anything I can do to make her enjoy life more? It seems such a pity when Sammi is so happy.

Yours Sincerely,

Monica McFarlane

The aim of being alive, the evolutionists tell us, is to live long enough to procreate and ensure the survival of our genes.

The ultimate aim of this process is unknown and almost too magnificent to contemplate. While we can trace the development of many life forms on the earth over millions of years through studies of fossils and traces left by species long since extinct, no one as yet has been able to create even the simplest species from a bottle of the right type of chemicals, nor has anyone been able to show why the survival of the fittest of each species, or even each species itself, is important. Not being a philosopher or an expert in divine purpose, I should perhaps confine myself to how each cat strives to survive and pass on its genetic complement to the next generation.

Human intervention, of course, denies the prospect of reproduction in most cases by neutering the cat, yet the will to survive obviously persists. The cat, like every other animal, is born with the capacity to respond to challenge and protect itself from the life-threatening dangers of attack by predators and environmental risks such as fire. The responses to such challenges are easily distinguishable from the cat's normal (most often expressed) behavior patterns. The startle reaction is genetically programmed, and even young kittens are seen to arch their backs, erect their fur, and flatten their ears when alarmed by a loud noise or sudden unfamiliar happening. During the early weeks of life a kitten, like many other mammals, will run to Mum for protection immediately after being startled if she hasn't already intervened. Gradually the kitten habituates to commonly encountered noises or environmental stimuli, especially if they are never actually followed by anything harmful or physically painful.

In short the kitten grows up constantly learning which stimuli are potentially threatening and which are not, yet without ever losing the capacity to demonstrate the initial startle response. In the average upbringing each kitten is challenged by a huge number of environmental changes and social encounters, first with its littermates and mother and later with other cats. It learns to adapt and take most things in stride. Once learned, its competence at dealing with a given stimulus without showing a fearful reaction is maintained by occasional exposure to it. Repetition of exposure, maternal teaching, and the development of exploratory behavior, which propels the kitten into new environments, all help it learn to deal with novelty and use those innate patterns of behavior to deal with anything threatening. Monica McFarlane's Sammi is a typical competent, confident cat that has gone through the feline school of life and, suitably manipulated by us, has become the ideal pet.

For others nervous responses such as seeking to avoid any challenge by remaining in a dark corner or behind a solid protecting barrier like a sofa, or low-threshold, over-quick flight reactions when experiencing slight changes to the environment, ensure that the cat never learns to cope. The anxious cat is easily identified by its crouching gait, low carriage of tail, and slow, low movements toward a sheltered spot under a table or in a corner. Once as secure as it can make itself, it may then face the world, still hunched and trying not to draw attention to itself, pupils dilated as a response to adrenaline, a hormone released at such times to prepare its muscles for running away. Alternatively it may adopt the ostrich method of avoiding challenge. Having found a secure bolt hole, it then avoids facing the challenge altogether and lies still, hoping for it to disappear. Some even learn to hide themselves under the bedclothes when anxious, similar to that famous "head in the sand" method of making worries go away. Whichever reaction the cat chooses, it is clearly distressed, and it is distressing for the owner to observe as well, especially if the cat reacts in such a manner in response to harmless, everyday events and never comes to grips with life. The signs of its distress are clear, yet the cat is difficult to comfort at such times without contributing further to its anxiety. At the very time it is trying to avoid challenges we draw attention to it, and it may never learn that our efforts are designed to comfort it. If we push too hard, it may even lash out as a last line of defense when there is nowhere else to run, and the threshold of its nervousness falls ever lower as a result. One such cat is Shelley, the opposite of Sammi in the competence ratings and clearly not enjoying life to the full.

Unhabituated Responses

Kittens brought up by inexperienced or incompetent mothers will be less competent themselves at dealing with change, as indeed will those relatively unexposed to challenge during the first weeks of life. Their survival chances in a wild world would not usually be so good. There is enormous individual variation among kittens' thresholds of reactions to different stimuli, even among individuals in a litter apparently exposed to the same events from birth. While young animals always habituate to stimuli faster than adults, the smaller or inherently less robust individual may hold back from its littermates in advancing to investigate a new toy because it has learned that it may get caught up or bowled over in the rush. It will avoid this risk by waiting until last or not advancing to investigate at all. As a result it will experience one fewer event than the others and be less competent by comparison. Faced with the same toy later, the other kittens will already have learned to overcome any reluctance to investigate and leap straight in to play or ignore it. By contrast the inexperienced kitten may show an initial startle reaction to the toy when finally encountered or, worse, never dare to experience it at all.

Perhaps Shelley is an example of the adult result of the reluctant and incompetent kitten. Such general nervousness is termed *unhabituated anxiety*—the cat simply received a range of experiences as a kitten inadequate to enable it to cope with the flow of challenges as an adult. Controlled exposure techniques can be designed to allow Shelley to experience what we and Sammi would regard as commonplace and nonthreatening events in the home and even outdoors. In short we can provide fresh opportunity for Shelley to learn to cope with all those happenings. However, in such cases it is rare to produce a totally competent, "normal" cat as there are always certain stimuli that can be habituated only as a kitten because of subsequent negative experiences or the failure to have positive ones. For example, kittens brought up with puppies, or even adult dogs, rarely show nervous reactions when dogs approach them as adults to play. By contrast the adult cat that has no experience with dogs is unlikely ever to be relaxed about such advances and usually retreats quickly. Some older cats can learn to live with a dog late in life, but it takes an inordinate amount of patience, and most get only as far as tolerating dogs and rarely will any social interaction or play develop.

In treating any nervous animal, progress can proceed only as far as its present capacity to learn allows in the light of its sum total of experiences. Improvements can be made with cats like Shelley by providing the cat with a new den (a kittening pen is ideal), which is placed in the main activity room of the house. Normal family life goes on as usual around the cat, which is protected from physical danger by the pen. Most importantly, the cat is also prevented from running away and avoiding the challenges of changes in personnel, movement, noises on the television, or movement of furniture going on around it. So it has to face them and start to interpret what is happening. The cat is essentially in a womb, warm and given all life-support systems of food, water, and, unique to wombs, a litter tray, but in a position to assimilate and habituate to stimuli that previously induced a fearful, avoiding reaction. The same techniques can even be used outdoors with certain agoraphobic cats (page 42).

Severe cases of unhabituated nervousness can also be helped with sedative treatment. Yes, cats with certain types of problem can be given Valium, as can their long-suffering owners. In treating most nervous conditions it is important that the sedative be withdrawn gradually as the cat's competence to deal with life improves with controlled exposure, so that its new competence does not become dependent on the drug. In short the drug is a vehicle for the learning process so that the cat gets the best out of its capacity to learn without being overshadowed by an unnecessarily pronounced initial fearful reaction. In some cases a low maintenance dose of sedative may be required long-term, or even permanently, for the cat to be more confident, but fortunately the majority of even severe cases need only a short course to accompany the exposure side of treatment. While I am no expert in alternative medicine, certain types of homeopathic or herbal treatment have also been reported to serve as effective alternatives to sedatives and are increasingly being prescribed by veterinary surgeons in certain cases. But, as with conventional veterinary medicine, it is vital that advice be sought only from a properly accountable practitioner of homeopathic veterinary medicine.

Unhabituated nervous cats will usually worsen if left untreated. The majority of such cases referred to me have concerned cats aged three years and more. They are often described by their owners as having been shy or withdrawn from the day they arrived, and their tolerance of the owner's lifestyle and home goings-on has steadily worsened. The threshold of their flight reaction falls to the point where almost any noise or quick movement is enough to make them run to a bolt hole, and stay there long after the problem has passed. Other typical signs include increasingly secretive, withdrawn behavior by the cat, which comes to avoid all change and take an ever-increasing time to come out into the open areas of the house when the owner returns after a period away. The cat seeks dark, quiet corners for shelter and spends long periods there, and may even venture out for food only under cover of darkness.

The picture is often a sorry one and highly distressing for the owner to watch, yet the signs of a worsening condition have been evident for some time. The longer the cat is left untreated, the harder it is to reverse the trend of withdrawal and to restore the cat to its original level of competence and beyond to greater things.

More rarely the unhabituated nervous cat may cling to its owner constantly for security and show marked and generalized nervous reactions when parted. Overdependence is particularly difficult to treat for the owner as well as the cat because it involves a loosening of that bond, but it should be treated for the sake of the cat's emotional state when away from its owner, if nothing else.

Work by behaviorists around the world has clearly estab-

lished the importance of frequent handling of kittens between the ages of two and seven weeks to habituate them to
human contact and lifestyle and make them into tolerant
and friendly pets. Essentially we overcome any initial startle
reaction by flooding the kitten with approach and contact,
which it soon learns is not threatening. It finds that such
handling is even enjoyable, being similar to the contact
provided by its mother. It also learns that our homes are safe
and that being in an open area in our home gets it noticed
and more likely to receive some of that comforting maternal
affection. The same is largely true of puppies, and by the
time most of our pets are weaned and becoming socially
aware they are long over any fear of humans . . . usually.

Loss of Habituated Responses

Dear Mr. Neville,

*I read about you in a women's magazine and
wondered if you could help with my cat. I should
say that I have had cats all my life and am glad
to say that none of them has ever sprayed, soiled
in the headphones of the hi-fi, or eaten my underwear. But Topper is now nervous about visitors
and always runs away when anyone calls. He
used to be very tolerant and even friendly, and
he's still okay with me and my husband, but I feel
so sorry for him being afraid to stay in his own
home when my friends only want to be nice to
him. As soon as the bell goes, he's off and won't
return until after they've gone. Is there anything
we can do? Help!*

Yours hopefully,

Joyce Hayes

Topper is a prime example of the cat that learned to cope
with the arrival of strange people in the heart of his territory

but then steadily lost the ability. I see many such cases where the cat is often otherwise untroubled by changes within the house, but it completely loses the earlier habituation to the stimulus of the arrival and presence of its owners' guests.

The converse of this problem is the frequently observed pastime of confident cats blessed with a sense of humor. They leap onto the lap of any visitor who dislikes or, even more unfortunately, is allergic to cats as soon as he or she sits down. The cat seems to enjoy the discomfort of the poor person, who is naturally unwilling to offend the host by pushing the cat off and instead shifts or sniffs uncomfortably while the cat dominates the proceedings.

But Topper's nervousness about visitors may not necessarily have been caused by lack of suitable experience with enough different people when young. It could be due to a single unfortunate experience with a particularly noisy or unkind guest that frightened him and taught him to avoid all risk of repetition in the future by running away early. Such single traumatic incidents also lie behind the development of many specific phobias in dogs and people, but I have yet to see a case of a phobic cat, other than agoraphobia, where the cat is afraid of open spaces indoors or, more usually, is afraid to go outdoors. Common canine phobias include pronounced fear of wind, rain, and sound, and I have even seen cases of dogs that are phone booth and light bulb phobic! Nervousness of any type can perhaps be described as more general or less specific than a genuine phobia, but the principles of treatment are based on desensitization, with controlled exposure of the stimulus in a diluted manner that does not invoke the fearful or phobic response in the patient. The stimulus intensity is steadily increased, and the patient habituates to it as a learning process.

Cats are natural avoiders of potential conflict as a survival mechanism and have highly developed senses to enable them to detect danger and respond quickly with flight if necessary. That they do not wait to investigate an apparently dangerous stimulus such as a car backfiring has helped to ensure their survival as a solo creature that cannot rely on fellows to protect it if caught napping. The difficulty in treating some cases of nervousness lies in determining what is "normal" for the cat that is judged to be "nervous" and whose quality of life is affected as a result.

Clearly Topper is not enjoying the potential benefits of meeting his owner's friends and regards them as undesirable threats to be avoided by running away. This is a perfectly normal reaction designed to ensure the cat's survival; it's just that there is no reason for the cat to perceive that its survival is threatened. The more guests the owner has, the less she sees of Topper and the less he regards home as a secure base. The end result is a worsening relationship with the cat, which may at all other times be thoroughly rewarding, or a change in the family's social life because they stop inviting friends around for fear of upsetting the cat. Either way there is a problem to be resolved, and in most such cases resolution is perfectly possible.

However, treatment is rarely simple and does not involve hanging on to a panicking cat while trying to press him into the arms of even the most understanding cat-loving guest. Being thrown in at the deep end doesn't always teach you to swim—it can cause drowning. In cases such as Topper's such action would be guaranteed to enhance his fear of visitors by adding the prospect of restriction by his owner to his apprehension, quite apart from any pain experienced from the owner's frantic efforts to keep a tight hold. Fortunately most owners try this approach only once and call me while their injuries are still fresh. Such cats usually calm down quite quickly once at a safe distance (the flight distance), groom themselves, and then get ready to run when they hear footsteps approaching the door, not even waiting for the doorbell to ring.

Treating Topper involves the same sort of controlled exposure techniques as when treating Shelley, but targeted more directly at overcoming the specific problem of accepting visitors. The first aim of treatment is to block Topper's attempts to escape or avoid exposure to the challenge. His

success in doing so, although protecting him from the
danger he perceives, also precludes any possibility of his
learning to cope on his own. Instead Topper is denied the
opportunity to avoid visitors either by being restrained on a
leash if he is comfortable wearing a collar or harness or by
being kept in a traveling cat basket for short periods when he
is to receive guests. This is placed in the area where guests
are invited to relax—usually the living room—before they
arrive. The more willing volunteers there are, the better,
although Topper should first receive guests that he knows,
such as members of the family. They ring the doorbell instead of using the key. Topper's first reaction is the usual one
of alarm and an attempt to escape, but this is prevented by
the basket. Then the guest enters and Topper, seeing that it
is only Joyce or her husband, quickly calms down. Repetition
should cause Topper to start to associate the doorbell with
nonthreatening arrivals.

Later less-well-tolerated guests can be asked to perform
the same routine, entering Topper's room with an accepted
member of the family and doing nothing more than sitting
down some distance from him. It is essential that Topper
grow used to their presence in gradual stages. Now his cage
serves to protect him from the challenge he has avoided for
so long, and he should settle quickly.

Slow progress at this stage can often be speeded up by
quelling the cat's overreactions with a little sedative treatment such as Valium by prescription from the veterinary
surgeon. However, as with cases of severe unhabituated anxiety, it is essential that the cat's tolerance not become dependent on drugs, and a tapered dose prescription is usually
employed. The tolerance of guests may be totally drug-
dependent for a few days, but as he is exposed to more
visitors the drug is slowly withdrawn so that his tolerance is
increasingly learned and decreasingly drug-dependent. The
drugs are simply a vehicle for exposing Topper to his problem. On or off drugs, with frequent exposure to as many
different people as possible under the right conditions,
Topper should perceive their arrival in and occupancy of the
core of his territory as being neutral. More importantly, they

are able to remain inside his flight distance, the distance that defines his opportunity to escape in the same way that zebras maintain a flight distance from the lions on the African plains. It is a fundamental distance in defining chances of survival in nature. Those that misjudge the distance they need to effect escape from any sudden challenge through illness, age, or incompetence get eaten first.

The next stage of treatment is a little more invasive. Now guests are asked to sit progressively closer to Topper's cage to habituate him further to their presence. This stage can proceed only as fast as Topper can tolerate, and guests should certainly not attempt to touch him or even talk to him until he seems confident about their presence. Then we introduce the prospect of forcing the issue a little. Though it sounds a little unfair, Topper should be starved for twelve to twenty-four hours so that he is hungry when pressed into sharing space with his next visitor. The visitor, sitting close by Topper's cage rather than bending down over it, which would alarm him, gently proffers a small tidbit or tasty portion of a favorite food through the bars of the cage. Topper's positive appetite-stimulated reaction to the food is designed to equal or be in excess of the intensity of his nervous reaction and so help him overcome it. Again this stage of treatment can be assisted by certain drugs from the vet that increase the cat's appetite and therefore his incentive to come forward to guests. The drug of choice, which interestingly is used more generally as a contraceptive for cats and dogs, also has a sedative effect and so may be used from the early stages of treatment in severe cases. Food cements relations far quicker than gentle voices, though the visitor and owner should encourage proceedings by talking gently to the cat while offering food. Thereafter the cat should be fed frequent short meals for the length of the visitor's stay (or patience), and as many guests as possible, as well as the family, should take part in the process. This steadily spreads the cat's confidence around and helps him view all guests as potential providers of food and, later, affection.

The final stage of treatment is to dispense with the cage

and restrain Topper on a harness or collar and lead when guests arrive and to have them offer food as before. Then the owner holds him as before and takes him toward a single known and accepted visitor. This should be done slowly so that Topper doesn't panic as he did before he learned that visitors could be nice. The advance should be slowed or halted if Topper starts to look alarmed or struggles. Once alongside, guests can gradually start to stroke him. This is most readily accepted if Topper is first stroked by Joyce, the guest slowly joining in. The process is complete when the owner stops and the cat is being stroked by only the guest. Holding Topper may not be possible for guests for some time yet, if at all, because this is an enclosing action denying escape and thus involving Topper's total confidence. He would only be acting in the same way as a great proportion of cats if he were to restrict this honor to his family only.

During all contact guests' hands should approach initially either from the side of the cat, unseen, or very slowly from directly in front of Topper so that he can see and accept the advancing paw. The hand may be very much a paw to Topper and should be offered very gently indeed, bearing in mind that paws are a cat's main armory and therefore likely to cause him to be apprehensive in the same way as he would be apprehensive of a swing from a cat's paw.

It may even help to approach Topper at the same level rather than intimidating him unnecessarily by bending over him. While the prospect of crawling along the floor to approach him may seem like carrying things to extremes, placing Topper—whether in the basket, on a lead, or hand held—on a table and approaching face to face could be less threatening. Of course all volunteers must be safe, and if there is any prospect at all of Topper lashing out defensively with a paw he should be in his basket when he receives guests for some time yet. Approaching face to face mimics the greeting behavior of cats, and as our face shape is not too dissimilar from that of a cat with regard to forward-facing eyes, it may be more readily tolerated than a big mitt landing on top of Topper's head. Obviously all volunteers should avoid making loud noises or sudden movements as they approach him, to avoid inducing the startle reaction; equally they should avoid staring at him as this could be seen as a challenging gesture similar to the staring "battle of wills" observed in cats facing off over territorial disputes. The nonterritorial Topper would always run to avoid that type of confrontation with a visitor, so they should look below his eye level or to one side as they approach him.

Progress with such cases is normally steady, and once improved the cat rarely regresses, provided he is regularly exposed to visitors. Essentially the methods of treatment are very similar to those with many other behavior problems. The opportunity for success of unwanted behavior or response is denied physically, the physiological reaction of fear or panic is muted by pharmacological means if necessary, and the animal is made to face up to the stimuli that cause it to be nervous, in a slowly presented, controlled fashion. Gradually the patient learns first to accept, cope with, and tolerate the stimulus and ultimately perhaps even comes to enjoy it. The key to success is to reduce all variables so that everything is controlled and all possible reactions from the cat are expected and to proceed in short, frequent doses of exposure at a pace that the cat can take without inducing panic. When not being exposed to visitors, the cat should be allowed free range as before but should be placed in its basket periodically without being confronted by visitors so that it doesn't come to perceive it as an unpleasant restriction. Ideally it should be fed in the basket at normal mealtimes to maintain the idea of the basket having a variety of uses, some of which at least are good.

It usually works. However, children and loud, bad-tempered, or drunken guests or disabled visitors may still be alarming to Topper after he has apparently improved, because of their unpredictability and less rational behavior. Given that controlled exposure techniques may be difficult to organize, he should perhaps be protected from such unusual invasion by being put out or in another room before they arrive. This will preclude any possibility of regression to his old fearful ways unless, of course, drunkenness is more the norm than sobriety, as was once the case of a cat I saw that had been rescued from a pub in Scotland. The cat, named Pernod, was very suspicious of people who were quiet and walked straight but had no difficulty at all in coping with staggering, loud drunks. All a question of early exposure, no doubt.

Agoraphobia

Dear Mr. Neville,

My cat Mugs is agoraphobic. He is very happy indoors and used to be happy outdoors, but for the last four or five months he won't go out at all. I've tried pushing him, but he just panics as we get near the back door and runs straight back inside if I push him out. If I shut the door, he just screams by the back door until I let him in again. So now he has a litter tray in the kitchen and never goes out, and I'm sure he's missing out on life as a result. Do you think he is suffering? Should I insist he go out?

Yours sincerely,

Jonathon Ringer

Agoraphobia is the fear of open spaces, which includes the outdoors but can also mean open areas within a room. Fortunately this is quite a rare condition in cats and is the only true phobia I have ever encountered in the species. The condition can arise from a lack of early exposure to the outdoors or exposure delayed too long and not coinciding with a kitten's exploratory phases of development after weaning. Sadly, from the behavioral development point of view, kittens cannot be allowed outdoors in safety until their vaccination course against unpleasant and often killer diseases is complete at about twelve to fourteen weeks of age.

Ideally kittens should be exposed to the outdoors as soon as they are weaned so that maximum competence at coping with the brave new world can develop, but clearly I would never recommend putting the kitten's health or life at risk to achieve this. Instead responsible breeders ensure that their kittens are exposed to as wide a variety of novelty, handling, and challenge indoors, in safety, so that they all grow up with as wide a range of abilities to cope as possible, which will include exploratory behavior patterns to help them cope with the outdoors as and when they are allowed access. Therefore my advice to prospective owners of a pedigree kitten is to seek out a breeder who is aware of the behavioral needs of kittens and who is not just raising money making cat bodies in an unstimulating pen at the end of the garden.

For such victims of callous breeders the prospect of being incompetent outdoors or even agoraphobic is a common development. It is perhaps surprising that so many kittens overcome such a lack of early stimulation to become well-adjusted nonagoraphobic cats, but for a few their unhabituated anxiety problems could so easily have been avoided.

The majority of the agoraphobic cats I treat are unwilling to go out or walk across the center of a room because of a loss of confidence caused by a single traumatic incident, usually a fight with a highly territorial local rival. Trauma or uncontrolled exposure is the more usual cause behind the onset of most phobias. The most common such trauma in the development of agoraphobia is when a fight with a rival occurs in the cat's own house, where it used to be completely safe. More often such fights occur outside, and one suspects that such is the case with Mugs.

The risk of encountering the rival may be present every time he goes out, and his reluctance to do so thus increases steadily. He may lose the ability to cope outdoors with even mild changes such as the sound of leaves rustling or the noise of a car, even if the other cat is not on the scene. Typically, the severity of the cat's reaction is pronounced with all exposures, even on apparently calm days, when no other cat can be seen. But Mugs's agoraphobia may be due to other causes, including major disturbance of access points to home base during the building of an extension or garage, a chance encounter with a stray dog in the garden, or a near miss with a fast car. The consequences in all cases are usually quite marked and get progressively worse. The cat is clearly distressed if forced outdoors and a long, long way back from enjoying its earlier outdoor lifestyle.

Treatment for cats like Mugs is similar to that for other forms of nervousness and involves controlled exposure to the outdoors, perhaps using the carrying basket again or, better still, a large, secure pen in which the cat can safely spend some portion of his day outdoors. Such treatment is usually best delayed until the cause of the problem is removed. This may mean waiting until building work is complete or even coming to an arrangement with the owner of the despotic rival about which cat is allowed out at which times, to avoid further conflict. Once achieved, Mugs can be put safely outdoors to relearn that outdoors is as safe as before the trauma and that every noise on the wind is not necessarily threatening. Jonathon should accompany him on the first few occasions and walk around the garden with Mugs on a lead so that he acts as a security bridge for him. Sometimes it helps to divide the cat's meals into frequent short rations and to move his feeding area to the pen and later just outside the back door to use appetite again as a distraction when the cat feels vulnerable.

The prognosis for treatment of cats such as Mugs is often very good, depending on the control that can be achieved over the cause of the initial problem, because we are reestablishing old habituation patterns rather than trying to teach them to the cat for the first time.

Husband Phobia

Dear Mr. Neville,

Our Persian cat, Fluffs, adores me, my children, and even our rabbit but hates my husband. The

*poor man has tried for months to get the cat to
like him but to no avail. As soon as he enters the
room, Fluffs runs away. John has never laid a
finger on him and only ever wanted to stroke him
and be friends. Is there a way to get them together? Fluffs is just over a year old and doesn't
go out because we live on a busy road. He was
neutered about six months ago, but this made no
difference to his dislike of my husband. It's really
becoming a problem now because my husband
gets upset that I devote extra attention to Fluffs to
make up for what he can't get from him. Please
help before we come to blows!*

Yours truly,

Karen Hayward

Dislike of husbands is more usually restricted to wives, they
tell me, and rarely if ever is described as a phobia! However,
over the last year or so I have received a number of letters
similar to that of the unfortunate Mrs. Hayward. Some cats
do bond very closely to one maternal figure but are usually
tolerant of other members of the family without doting on
them in the same way. For a cat to reject one particular
individual while remaining friendly with every other creature
in the house, including noisy kids and tasty prey items such
as rabbits, is unusual and a real blow to the often innocent
victim. As in this case, the victim is usually the husband. As
in this case, the husband has tried extremely hard to overcome the cat's dislike for him by trying to be ultracalm and
affectionate. Sadly the efforts have failed, and now other
relationships in the family have begun to suffer because of
the competition for affection from the cat and also between
the accepted family members and the person rejected. It
may seem strange to begin to quarrel over a cat, but being
rejected by any member of a social group is bound to hurt.
If it happens to be a cat that you love dearly, treat very much
as one of the family, and spend much time and money on,

then it is hard not to take the hurt personally or resent the success of others who perhaps expend less emotional effort in being nice to the wretched cat.

Why cats should be more likely to be nervous about the man of the house rather than any other member is difficult to say. In some cases the man is obviously short-tempered and prone to crashing around and shouting when upset. This is likely to upset the cat, which will be unable to feel confident in his presence, even when he is relaxed and pleasant.

Yet in other cases the wife appears the more volatile and the husband is the calm, constant, and loving influence in the household. Perhaps then we must consider the early experiences of the cat and investigate who took the major role in feeding, being affectionate, and playing with it to understand whether the bonding process was shared a little unequally. Certainly for some husbands there is no obvious reason why their cat should dislike them. It may be that the husband simply isn't at home as often as the family and that his arrival sparks off much sudden family movement when he is greeted upon arrival from work, which the cat finds startling and so keeps out of his way.

I am told by a source close to home that one particular man may be likely to upset the resident cats at any time because of his general unpredictability. This man is apparently prone to great swings of moods, being calm and pleasant one minute, then banging the table and making loud roars of disapproval on receiving some minor piece of unimportant news, then, with emotions released, returning to his normal calm self. "Small wonder," says this informant, "that cats might find men unpredictable or even undesirable and learn that keeping out of their way is a better policy than constantly trying to anticipate their moods." But as both my cats like me, this cannot apply to the case she cites!

Treatment of husband-phobic cats has to be a family affair, and often this necessary corporate approach immediately restores the strained human relations to their former strength. The common aim of helping the cat draws the family together, given the right explanations, even though the advice is not always too pleasant for the wife to carry out. She and other closely appreciated family members are advised to be far less available to the cat and to reject its demands for attention for a few weeks. Meals are divided up again into short feeds available exclusively from the husband, who is also the treat provider and the one who opens the door to let Fluffs in and out. Games are begun by the wife but are taken over by the husband, and he is allowed the armchair closest to the fire or radiator (if he hasn't taken it for his own already) so as to be more attractive to the cat. He is advised to be more in control of his emotions and to present a more consistent appearance of calm and friendship for the cat.

The relationship usually builds more quickly if the husband gives up trying to pursue the unwilling cat and waits for the cat to need him, while providing a few incentives. And big chaps especially are advised to get down to floor level to mimic those feline greeting behaviors rather than lean down from on high, which can be frightening for the unsure cat. All fast movements and loud expletives are out too. Progress may be quite slow in some cases but astoundingly quick in others, and for no apparent reason. In particularly slow cases I have even suggested that the husband stop wearing after-shave or using "stud" soaps on the premise that many such "masculine" toiletries are scented with real or artificial male animal extracts such as musk oil and therefore present an unnatural challenge to the cat that we could never perceive.

The most rewarding letter I ever received after treating a husband-phobic cat explained briefly that the cat was much better and would sit on the man's knee to be stroked, and even occasionally purr, but best of all the husband was a totally reformed character who had given up all signs of macho, short-tempered, and dominating behavior in the home for fear of undoing the progress and upsetting the cat. And for all that they got only a small bill for the cat's treatment!

Nervous or phobic, insecure or lacking in confidence,

the cat makes a good patient for the cat shrink because of its intelligence, adaptive learning capacity, and highly developed individual survival instincts. Owners too are usually compassionate and more than willing to invest the necessary time and management to carry out treatment suggestions. Usually their efforts pay off and the cat's quality of life and their joys in owning him or her improve steadily.

Bonding Problems

Dear Peter Neville,

I have the most wonderful cat in the world. His name is Baby, he is four months old, and we love each other dearly. He wants only to be with me when I am at home and follows me constantly. He sits on the edge of the bath when I'm in it and waits outside the door when I'm in the loo. He'd follow me there if I let him. He's quite the most affectionate cat I've ever known and loves to be cuddled. He must think I'm his mum, because he purrs loudly and sucks on my clothes as if they were nipples. I don't mind this at all except when he decides that clothes aren't good enough and crawls up to my neck and sucks it. Even though I have red rashes from his attentions, I haven't the heart to put him down as I would hate for him to feel that I was rejecting his advances. Is there anything else I could do?

Yours sincerely,

Mrs. Beverley Moore

Baby is an entirely appropriate name for this loving young cat who avoided the traumatic experiences of being isolated from his real mother by transferring all infantile dependence to his new human owner. It's simply a case of that maternal relationship we would have anyway with the cat, as described in the first chapters, being allowed to continue to the same pitch with the owner. Although Baby has made the nutritional break of independence from his mother and eats solid food, not milk, the normal process of physical rejection and isolation from the caring, nurturing parent has not been completed due to the availability of a more than willing substitute. The result is that infantile behavior and dependence on a maternal figure continue long after they should have been replaced by more adult patterns. The dependence on the mother for protection is maintained with the owner by following her everywhere so that Baby can reach her quickly should danger threaten. Normal exploratory behavior and the development of independent decision making is markedly slowed, often to the point of incompetence or the development of nervous conditions when the cat cannot rely on its owner for security. The cat may even start to suffer from separation anxiety when the owner is away, a problem more usually encountered with the pack dog for which isolation is unnatural. For a creature that is largely an independent, solo hunter when adult, separation anxieties ought never to occur. That cats such as Baby often fall to pieces and hide themselves away in a dark corner when left alone is most definitely to be avoided.

However obvious the treatment of this case may be, Mrs. Moore should be tackled gently. She is only doing what all cat owners do, enjoying that maternal relationship of being a provider of food, love, and protection. It's just that she is probably being too conscientious with a young cat that is quite happy to continue being a kitten. The rewards of the kitten not becoming a cat are that it also retains much of the sporadic play behavior we find so enchanting and demands our attention, which is in itself flattering and enhances our affectionate feelings.

Detachment may be long overdue for the sake of Mrs. Moore's neck as well as Baby's development, but explaining the procedure without causing Mrs. Moore to feel that she is rejecting her pet is a delicate affair. In this situation counseling skills may be far more important than any knowledge of cat behavior. Mrs. Moore should become far less available to Baby, and affection should be initiated by her rather than by Baby. He should be prevented from following her and encouraged to spend more time developing new interests outdoors and given more opportunity to explore indoors. Mrs. Moore's family should take over the caring duties of feeding and grooming Baby to help him spread his loyalties more and be less focused on "Mum." Most difficult of all, Mrs. Moore has to offer far less affection and only in short doses, designed to mimic greeting and social bond-maintaining contact as found between friendly adult cats. Prolonged nursing-type cuddling sessions should be stepped down steadily and never allowed to reach the point where Baby relaxes sufficiently to begin sucking on Mrs. Moore's neck!

Baby will grow up quickly and become more independent, exploratory, and competent to cope with life away from the apron strings. Mrs. Moore may have to be helped to overcome feelings of guilt during the process, but my role is to help her justify the rejection and equate it in terms of allowing Baby to become a young boy without running the risk of losing his affection altogether. Perhaps a case of worrying too much, but a situation we can all unwittingly find ourselves in with even a well-adjusted adult cat if he happens to fall ill or get injured.

Intensive Nursing

Modern veterinary treatment often allows us to nurse what were previously hopeless cases of injury or sickness in cats, with a good chance of recovery. While the latest drugs counter infection and the latest surgery resets bones and grafts skin, we as owners become the attentive nurse once the cat is allowed home. For days or weeks, and sometimes months, we tend our poor pets through all hours, feeding liquified food through syringes and massaging limbs to keep them ready to function when our favorite cat starts to get about again. The cat can make an amazing recovery against all medical odds, but afterward it can be something of an emotional wreck. The problem is that over those weeks of nursing it has come to depend totally on its owner again for food, comfort, social contact, and even to clean up when nature has called on the recumbent immobile cat. The psychological result is a recovered but overattached cat who has become as dependent on its owner as when ill or as a helpless kitten. This condition is fortunately not too common as the cat becomes increasingly independent again as it recovers, but for some owners the aftereffects of injury and illness are as tough and demanding to deal with as the nursing itself.

Detachment once again is required and in exactly similar fashion to that from the overattached young cat but, again, strikes hard on the owner who has grown used to worrying about the cat's every need during its nursing. Good counseling skills are required to help the owner comprehend the need to ignore his or her pet's screams and wails for attention, desire to be hand-fed, and constant efforts to follow the owner. But the owners don't always thank me for the basically unwelcome, if effective, advice. Anyone know of a consultant in feline behaviorists' behavior specializing in rejection and anxiety problems?

Leaving Home

Fortunately there is the other side to the attachment coin, the side where my advice is to increase the volume or frequency of the contact and affection on all fronts to help a distressed cat cope with life or to enhance the bond with its owners. Baldrick is a cat who was feeling the pressure of his owners' hectic lifestyle and found life in their home increasingly unattractive. His owners worked shifts, and he never knew who would be at home at any given time of the day or night. Feeding times were erratic or soulless affairs of a cold plate of food or just dry food left out by the owners as they passed through the house. A policeman's lot may not be a happy one, but this policeman and airport worker's cat's life was even less happy! Wild parties were frequent events when the owners' days off coincided, and poor Baldrick had to cope with the contrast of having a quiet house to himself one day and 3,000 watts per channel all the next night.

Yet occasionally one or other owner stayed at home all day, for the most part in bed, catching up on much needed rest, and then Baldrick was able to join the warm communal sleep heap. It was probably only those occasional opportunities to get together again that kept Baldrick around at all. Nonetheless he steadily spent more and more time away from home and grew distant from his owners. It was the summer of 1989, a long, hot summer in England when many cats were happy to spend far longer outside, sleeping rough and hunting to feed themselves. I received a great number of calls that year from owners worried about their cat's long excursions away from home. Of course, as the autumn approached and the weather became colder and wetter, their cats dutifully returned and spent more time at home, much to their owners' relief. It was a reminder for me of just how tenuous the relationship between man and cat can sometimes be and just how readily the cat can return to its ancestral wild ways and survive very well without us.

Baldrick didn't go all the way, however. Instead he set up camp at the end of the yard under a lean-to next to the greenhouse. Sheltered, protected from drafts, and warm on a pile of old newspapers, Baldrick learned to fend for himself and hunted for days on end in nearby fields. Yet he would also pop indoors from time to time to nibble on the best-quality free meals left by his owners sometime previously. Once in a while he would come in to sleep a day away with one or the other, but the owners noticed a developing reluctance by Baldrick to be handled. It was taking an increas-

ingly long time for him to recognize them if they happened
to come home while he was indoors. His first reaction on
their arrival was to run, and only sometimes could they call
him back. But what really made the owners decide to act was
the fact that Baldrick was starting to look like the grubby, ill-
kempt "Blackadder" BBC television character who gave him
his name. Baldrick the cat was a long-haired moggy who
became more and more matted and flea-infested as the
summer wore on. His increasing reluctance to be handled,
or even be near his owners unless asleep, was never so
marked as when they tried to groom him.

The situation had gone too far for gradual modification
of Baldrick's behavior. I advised quick, ruthless methods.
First the cat flap was set to allow Baldrick in but not out
again. His favorite food was placed in a feral cat trap, a
wonderfully ingenious device designed to catch unhandle-
able feral cats without hurting them and keep them unhandled as they are transferred to an equally ingenious
carrying basket with a movable side panel. In this Baldrick
was taken to his vet, squeezed gently to the side of the basket
by pulling the panel across, and suitably sedated and then
anesthetized by the vet. Once unconscious, Baldrick was
deloused, deflead, dematted, and thoroughly cleaned up and
then put back in his basket to be taken home to recover. He
was kept in a kittening pen in the front room for a week and
allowed out into that room only when one of his owners was
home. They adjusted their shifts to give him as much company as possible, and we tried to establish regular feeding
times for Baldrick around their schedules. If one could be at
home within an hour or two of feeding times, Baldrick would
be fed by them. If not, we employed another cunning piece
of technology, the timed automatic feeder. Food is placed in
a dish and the lid closed, timed to pop open and allow the cat
access to its dinner at its usual time. I had one once for my
cat Bullet; unfortunately my faithful dog destroyed it in
seconds to get at that most prized of foods.

The aim with Baldrick was to rebond him to his owners
with a concerted effort over the week's confinement through

frequent feeding, handling, and playing and as far as possible
to reassociate his owners as providers of food at regular
times. When they were not home, the food would still be
available only at those times. Hence the cat would always
have a vested interest to return then, and the owners were
usually guaranteed to find him in at those times. All meetings were thereafter to be effusive and, as expected, the
more the owners invested in trying to react with Baldrick
and be nice, the more he responded. Within a few short
weeks Baldrick was part of the family again, though he is
packed off to a neighbor's house for the night when all those
strange airport workers and policemen arrive to play loud
music, drink, and not drive home!

The Insecurities of Old Age

While Baldrick is an example of adapting a detaching cat to
a strange modern lifestyle with no routine, sometimes it is
we who must adapt to the cat's changing demands on us.
This is likely to be true of all cats as they age, and with cats
now living much longer we are starting to observe behavior
changes similar to those in elderly people. Instead of the
whole machine packing it in, cats and people alike wear out
a piece at a time. The slow wind-down or failure of an organ
or joint brings inabilities to respond in the same way as we
and our cats did when young. This in turn causes worry in us
and an increasing awareness in cats and people of a failing
ability to protect themselves or avoid potential hazard. For
all cats old age is a time for longer sleep periods and more
time spent safely indoors, and for many it is a time to give up
a little of that independent spirit and rely on one's benefac-
tors for emotional protection as well as physical care.

Dear Mr. Neville,

*Tiger and I have been together for over fifteen
years, and never have I suffered a single problem
with her. Recently, however, she has begun to*

*call pitifully at night. I get up and try to feed her
or sometimes simply cuddle her for a while and
put her back on her favorite blanket by the radiator. This stops the crying. My vet says she's as
healthy as can be expected of a cat her age and
can find nothing wrong with her. Why does she
cry out? Is she in pain?*

Yours, tired,

Anabelle Whitehouse

No, Anabelle, Tiger isn't in pain. She's calling out because
she needs the reassurance of your company. With many of
the oriental breeds excepted, cats communicate with people
and maintain their bonds of friendship through body language and the development of a communal smell through
physical contact. Vocal communication is largely one-way
from us to the cat, except during the first greeting or at
feeding time, when the cat will often issue forth a trill of
delight. Queens call loudly when in season to attract males,
but as most pets are spayed this is really suffered only by cat
breeders! The most unforgettable feline voice is that of a
constantly yelling Siamese queen in season, as many who
thought it would be nice to let her have a litter will testify.
Purring is rather different, and full meows and yells are
restricted largely to the Siamese, Burmese, and other Eastern races.

With increasing age cats are not only less active and
more dependent on us; they are also physically less able to
get to us as required. Needing more support to feel secure,
they learn to call to attract our attention. Annoying perhaps,
but the cat is most likely to feel insecure at night and periodically wake up and call out for reassurance. We are aggrieved
but always get up to see if the old boy is having trouble. Once
reassured that we are around and protecting, the cat falls
quiet and may be happy to go back to sleep. Reassured for
the moment, it has also learned that vocal communication is
the main method to be employed if feeling insecure and

continues to do it for the rest of its days. At this stage the owner is probably already preparing smaller, lower-protein meals for the cat and allowing it to sleep in previously off-limits areas because of its advancing years, so it's not usually too much of an imposition to allow the cat to sleep in the bedroom at night as well. It should be happy and comfortable, and at least if it does cry out and wake you up you can reassure it without getting up!

Routines of feeding and contact, as with Baldrick, can also help to maintain a solid, dependable relationship with the cat, and a secure bolt hole, warm, quiet, and out of reach of dogs and children, will be gladly accepted by the older cat. There are few places better than an old blanket on a shelf in the airing cupboard for this. Many cats teach their owners that trick along with that of the nighttime calling.

�20 6 �20

Stress and Trauma

Stress is a little-understood word that has so far been avoided in this book. Everyone knows what is meant by stress, and virtually all of us suffer from it at some time in our lives, if not constantly, yet we know little about it. Cats, too, can suffer from stress, and many of the problems I treat could be said to be reactions to stress of one sort or another. The interface among environment, behavior, and physiology is bared when we or our cats suffer from stress, and scientists are only slowly unraveling the complexities of it all.

Startle reactions and developmental learning help the cat cope with challenges and stay alive. Discretion is often the better part of valor for the cat, which will run away from danger quickly if the threat isn't discouraged by its startled arched-back posture of defense and warning hiss. More assertive individuals may actively repel living threats; a common example is the cat that has learned to deal with overfriendly dogs by standing its ground and swiping that cold, wet nose with a paw full of sharp dissuaders rather than running away. More nervous individuals may hide in safe corners or under bedclothes if unable to cope. Whichever method is adopted, the behavior changes are designed to reduce the conflict of the moment.

Physiologically the cat is able to respond because adrenaline is released into the bloodstream from the adrenal gland, situated above the kidney. Adrenaline primes the cat for fight or flight by causing the heart to pump faster and send more oxygenated blood to the muscles, while at the same time preparing the body to deal with increased carbon dioxide waste, which will result from the activity of running away or fighting off the challenge.

Stress is a very common cause of behavior changes in cats. Situations that can lead to stress include alterations in owner routine and lifestyle, retention in an unfamiliar environment, such as at the vet's office, or being locked in a cupboard accidentally. Overcrowding and prolonged exposure to worrying stimuli such as high-frequency sound can induce stress, as can grief following the death of a companion, human or feline or, occasionally, even canine. Cats can also find it stressful if, when with other cats, their flight distance is not maintained; this is the distance that they need to be assured of escape should trouble start.

When our behavioral responses to challenge don't succeed in reducing it, the adrenaline continues to flow to keep the body prepared. Stress is often the result and in human terms is manifested by irritability, tiredness, anxiety, and depression. Heart disease and stomach ulcers can result from the very mechanism designed to help us cope with the risk of injury if we are unable to avert the stressful challenge. Many of us adapt to stress and, in coping with the rigors of, for example, sitting in a city traffic jam every day en route to work, toughen up both physically and physiologically as our nervous systems evolve to be less sensitive to the adrenaline coursing through our bloodstream. This toughening up is comparable to the learning habituation to stimuli described in Chapter 4 in the physiological sense and is a vital process in the adaptation to changing environment for any species. If the stress of moving from favored forest shelter across an open plain was too much for a deer to the extent that it stayed in an old area with insufficient food rather than risk moving to a fresh area, then it would probably die out as a species quite quickly. That it can overcome the stress of moving across open land, where it is temporarily more vulnerable, by physiologically adapting to its body's usual responses to such challenge enables it to make the move and survive without suffering any harm from the stress itself.

Some people clearly enjoy being constantly in a state of readiness and, although stressed, do not suffer from any physical or psychological consequences. The difference between those who revel in such a state and those who are anxious, insomniac, or get stomach ulcers seems to lie in the biochemistry of their respective nervous systems and their individual sensitivity at the neurotransmitter level, rather than in a failure to cope due to lack of exposure when young or through being overloaded at first exposure, as when dealing with nervousness toward a particular finite challenge. Such an explanation, if true, at least helps us understand why everyone differs in the ability to cope with stress and, of course, why each cat will also vary.

While there is no evidence of stressed cats developing stomach ulcers or heart disease, clearly many do exhibit behavior changes and responses under stress that are similar to our own. The effectiveness of the immune system at combating infection may also fall as the prevalence of certain types of white blood cell may drop dramatically with prolonged exposure to stress in people and cats. Low-threshold aggression or irritability, depression manifested by secretive inactive behavior, and prolonged anxiety reactions all seem common. The appropriately named catatonia or "playing dead" is seen in stressed cats and other animals and is probably designed to stop the chase response in a predator, prevent the attack, and so help ensure survival. Indeed feline responses to stress of any kind are similar to those of other mammals and generally fall into one of two categories: excited reactions or inhibited ones.

Less frequently reported in the cat and perhaps only observable in the artificial confines of the experimental psychologist's laboratory are convulsions, hysterical epilepsy, excessive salivation, panting, colic, and even depigmentation of the skin (literally turning white with fear). The American veterinary surgeon and pet behaviorist Dr. Bonnie Beaver also lists anorexia, pronounced sensitivity to touch, changes in taste preferences, diarrhea, hair loss, and even psychological neutering, an alarming prospect with which I am thankfully unfamiliar in UK cats as most of them have already been surgically neutered anyway. Additionally there are some peculiarly feline responses to certain types of stress, such as sweating from the foot pads, urine spraying, and associative marking by scratching, urinating, or defecating on socially or geographically significant areas.

Alleviation of any stressful influence when the cat's reactions have failed to achieve their aim is one of the main features of treatment of many behavior problems referred to me. Identifying those influences often involves more than a little detective work. They may be environmental, if the cat's security in the house is not defined and it therefore lives constantly under the stress of being competed with for the home resources of food and shelter. They may be social, if the cat does not get along with other cats or one or more of the people that it has to share its house with. But whatever the cause, treatment of stress in the cat is largely achieved by removing or modifying the stressful influences or controlling the cat's exposure to them physically so that it can toughen up. Most cases involve a combination of both approaches, and this treatment is usually far more successful than attempting to diminish the physiological reactions of the cat by using stress-relieving tranquilizer drugs that block the activity of the neurotransmitters. While the stressful responses may disappear for the duration of the prescription, if nothing is altered when the treatment is withdrawn the cat will become stressed again and the behavior problems soon recur. The temptation is simply to keep the cat on drugs constantly to prevent stress reactions, but this is obviously only masking signs and not coming to grips with the real causes. However, used properly, tranquilizers can help in the treatment of stress in cats and, interestingly, are usually prescribed at about three times the dosage necessary to induce the same effects in man. But many cats respond quickly and well if the stressful influences can be identified and successfully manipulated without using tranquilizers long-term.

Psychogenic Dermatitis

Dear Mr. Neville,

James is a very sensitive cat. He is now four and was neutered when about six months of age. I have no other cats because much as I would like one I don't think James would ever accept another one and would be jealous. He is generally rather a nervous little soul and, although he enjoys going out, will often spend several days inside afterward if he has a scrap with another cat. Sometimes he will not eat for a few days and simply hide himself away if I have friends to call. He can also be rather snooty at times and ignores me if I have been away for a while or won't respond to his demands for affection. While I love him enormously and can accept that he is a very sensitive cat, I am worried because when many of these problems coincide he will nibble at the base of his tail and up his back. Subsequently he has flaky skin that requires veterinary treatment. My vet feels that the problem is psychological, and so I am writing in the hope that you may be able to help.

Yours sincerely,

Helena Bradley

Psychogenic or "nervous" dermatitis is probably a more common feline reaction to stress than we realize. "The

problem with skin," as a veterinary friend once said to me,
"is that so many things can cause the same problems, and
we often end up treating only the symptoms and hope the
cause goes away on its own." Now that there are veterinari-
ans with a special interest in dermatology, a few more of the
causes of eczema and dermatitis are being unraveled, stress
included. Diet sensitivity, particularly to preservatives, anti-
oxidants, and artificial flavors and colorings found in many
canned diets, or lack of certain fatty acids, are suspected as
causes of self-induced skin conditions by Mr. David Shearer
at Bristol University Veterinary School. Allergy (especially to
fleas) and postwound overgrooming can all set up skin
irritations such as dermatitis and miliary eczema. I suspect
that many stress-related conditions that owners do not consider serious enough to warrant veterinary attention also
arise, and so they go unrecorded.

The normal adult cat spends over a third of its waking
hours grooming to remove dander, parasites, mats, and
loose hair, which possibly also reduces the risk of parasite
infestation. Grooming helps the cat lose heat in hot weather
through cooling evaporation from licking and, most important of all, keeps the skin in good condition. It also helps
enforce social bonds between friendly animals that groom
each other. Indeed much of our relationship with the cat
centers on its acceptance of our groominglike activities of
stroking and tickling. The tactile self-stimulation of grooming by licking, scratching, or mild biting is also therapeutic
in helping to relieve tension, as we too experience through
scratching or combing ourselves when nervous.

A cat will often suddenly start to lick itself for no apparent reason. Perhaps it is responding to a minor irritation of
misaligned fur or, as is often seen in the face of obvious
stressful influences, as a diversionary activity that is more
comforting than facing up to the problem. Cats such as
James, which are sensitive to change or stress, or those
constantly challenged by unresolvable threats of one sort or
another may groom more often and more rigorously in an
effort to avoid exposing themselves to the perceived risks.

This nervous grooming can be the cause of psychogenic dermatitis and alopecia.

I tend to see cases of dermatitis only after all potential medical causes have been eliminated and the possibility of psychological influence is all that remains. Often those cats are very similar in general disposition to James, and, as with nervousness problems, treatment is usually a combination of modifying the cat's exposure to the stressful influence(s) as far as possible while providing an opportunity to learn to cope under controlled circumstances. For many a short period of hospitalization is a sufficient distraction to stop overgrooming and allow the essential treatment of the skin condition in a protected and nonstressful environment. In fact the slowing factor in treating psychogenic dermatitis is often the treatment of the dermatitis itself, which can be severe and, if it involves painful, crusty ulcerations and secondary infections, can take ages to heal. In the process the cat is responding to the itch by scratching and grooming the affected region, sometimes to the point where it has to be restrained from doing so by fitting an "Elizabethan collar." Treatment of the original psychological cause of the dermatitis is often assisted by a tapering-dose prescription of sedatives or tranquilizers, similar to that used in treating more general nervous conditions.

The scratching behavior may persist and maintain the condition long after the psychogenic cause has gone or the cat's competence has been improved. This can bring about a sort of self-reinforcing dermatitis that flares up when a series of stressful events arise. While we can alleviate intermittent bouts of stressful reactions using drugs short-term, such cases are usually dealt with perfectly well at the vet's and are not persistent enough to warrant referral to me.

Self-Mutilation

Self-mutilation is thankfully very rare in cats and, in both cases I have seen, has been a continuation of psychogenic overgrooming in response to very severe stress. Having set up an eczema reaction on the skin, both cats continued to lick and scratch at the affected areas, past the ulcerated skin and down through their own flesh. Though thankfully still rare, it is more commonly seen as a reaction to stress in dogs than cats. Self-inflicted damage to the flank area is an inherited disposition in many strains of Doberman, particularly in America. Many Dobermans elsewhere, my own dear Colonel included, will suck their own flank in an uncomfortable position when resting or worried. Flank sucking will often cause eczema to develop as a result of the skin's being frequently damp, but few cases go as far as self-mutilation down through the flesh. In cats or dogs it is clear that the pain of self-mutilation is overridden by the activity itself as a reliever of the stress the animal is under, yet the sufferer remains just as sensitive to other painful stimuli. Powerful nervous blocks that prevent pain avoidance responses must therefore be released in the brain or central nervous system during self-mutilation. This in turn indicates how strong the animal's motivation to avoid the stress of solitude, enemies, or other influences must be at the time it begins to mutilate itself. It appears that at the neurological level self-inflicted damage is preferable to merely the risk of damage from an outside influence. The continuation of the behavior is probably connected with the brain's release of natural opiates as relievers of pain when the cat or dog is injured, masking the sensation of pain to such a level that the therapeutic angle of self-interested, self-directed behavior continues. It is here at the microconcentration level of neurological chemistry that the best hope for treatment lies. At present we can only try to remove stress and offer usual sedatives in treatment, although the use of the morphine antagonist naloxone is being investigated in studies on horses in America and cats in Holland and may be researched at my Bristol University clinic in the future. Perhaps the first case will be one reported in mid-1990 of a Burmese cat that periodically tries to claw out its own tongue.

At my Bristol clinic I look forward to seeing cases of

psychogenic dermatitis and self-mutilation with Mr. David Shearer, who has a special interest in skin conditions. Hopefully between us we can come to grips with these problems using a combined veterinary/psychological approach to treatment.

Psychogenic Vomiting and Regurgitation

Dear Mr. Neville,

I have three wonderful moggies, two girls and a boy, all neutered. All are very different and have their likes and dislikes, but all are very affectionate with each other and me. The problem concerns the boy, Jasper, who vomits quite frequently when indoors. It seems to happen after the cats have had a small scrap or when I have visitors. My vet has tried feeding different diets, provided at different times of day, but to no avail. Could it be psychological? Is Jasper a compulsive vomiter, or is he just showing his dislike of some of my friends by being sick when they arrive?

Yours sincerely,

Frederick Reynolds

Many of us feel physically sick when frightened and when under stress of one sort or another. Fortunately few of us ever actually vomit except under severely stressful circumstances, and the same is true of cats. While most cats vomit occasionally or regurgitate hair balls with partially digested food, few are like Jasper. Such cats are usually rather incompetent generally and, although not giving out other signals of being particularly nervous, live their lives under continual stress because they never habituate to what other cats regard as normal occurrences. As a result it takes relatively little to force them over the threshold of stressful reaction and a single regurgitation of stomach contents at the time of peak stress, such as when visitors arrive in Jasper's house, or repeated retching and vomiting may occur as long as the stress persists and the cat is unable to avoid it.

Treatment involves steadily building up Jasper's confidence, as described in Chapter 5 on the treatment of nervous cats, and trying to offer specific controlled exposure to those challenges that tip the balance. Such exposure ideally occurs when the cat has an empty stomach, so management of feeding times and the provision of light, easily digested meals that pass through the stomach quickly are also features of treatment. In Jasper's case it may just be that he can never settle in the company of other cats in his home territory, and so the problem may be better managed by low-dosage sedative prescription if placing the cat in a new home is not considered an option by his owner. Where there is a prospect of such prescription being long-term, I am becoming increasingly keen to replace the use of addictive sedatives such as Valium with safer alternatives such as homeopathic or flower remedies under prescription or advice from veterinary surgeons with a special interest in these fields. The results with such cases so far have been very impressive, though the scientist in me still wants to know why. The important part of treating cats like Jasper is to produce some quick initial progress before Frederick runs out of friends who are less than impressed with Jasper's welcome, especially if they have come for dinner. The sound, sight, and smell of cats being sick is perfectly designed to put you off eating and ought to be marketed as an aid to human dieting programs!

Projectile Vomiters

Dear Mr. Neville,

I love cats, and I am the proud owner of three, all different breeds. Ming, my Siamese (neutered male) is always sick at night. But it isn't just a case of being sick, as when he throws up it is very liquid and there is lots of it. It comes out at very high speed and usually lands on the wall next to his bed. He doesn't seem to be in pain, and my vet can find nothing wrong with him, but it happens nearly every night, and I have to either get up and nurse him or leave it running down the wall until morning. I can honestly say the last thing I ever need to see first thing in the morning is cat sick. Is this a psychological problem? My vet has tried his hardest with everything from antibiotics to a pyloromyotomy [a loosening of the pyloric sphincter to ensure that food is not restricted in the stomach and can pass on normally to the rest of the gut] and suggested I write to you in case you had heard of it before and could offer any help.

Yours sincerely and rather tired of cleaning up,

Josephine Parker

No doubts about the diagnosis with this one: a projectile vomiter rather than a psychogenic vomiter, and a proficient one as well. It was difficult to know where to place this problem as it could equally well appear under the "Bizarre" section in Chapter 10, but since I treat it as a stress reaction due to lack of other sensible approaches from the behavioral point of view, here it is. Projectile vomiting can arise from a range of medical difficulties but in this context is usually reported only in Siamese cats. Typically the problems start at around six months of age; then the vet exhausts himself for a year with drug treatments and surgery in trying to cure the cat and finally gives up. At which point the cat may head in my direction, usually at one to two years of age and fighting fit!

Projectile vomiters are starting to get a strange hold over me as I find the behavior irresistibly fascinating. How can it be that an apparently healthy, competent, and sleek cat can have such a persistent and dramatic condition? Sometimes these cats are even slightly overweight, despite daily vomiting of at least some of their intake. Typically the vomiting occurs at night, four to five times per week, and usually after a period of rest or sleep. While a few are single cats in the home, the majority come from two- or three-cat households. In one recent case vomiting occurred so regularly that you could set your watch by it: four o'clock in the morning. The cat was heard to retch for a few seconds with that amazing regurgitative noise that only cats can make when being sick, followed by a huge ejection of the partially digested and often very liquid food at high velocity. It usually landed on a wall with a splat. Not a pleasant sound.

I now like to see such cases with Hilary Hill, a well-respected veterinary surgeon who was the Feline Advisory Bureau scholar at Bristol when I started my clinics there and is still on the staff. Like me, she finds projectile vomiters extremely interesting, and we continue to liaise over treatment of vomiters past and present. Many projectile vomiters are treated successfully by their own vet or by Hilary's expert attentions and surgery. She has found that some are caused by the overgrowth of certain types of bacteria at certain points in the digestive tract and can be treated successfully with appropriate antibiotics, but a few like Ming are unaltered by any such medical attention. I always refer these cases initially to Hilary for a full checkup in the light of her experiences. A few filter through to the cat shrink as a last resort. My success rate is 50 percent in achieving a reduction in the frequency of the vomiting, but although this sounds reasonable I should say I've seen only two cases, and we're not much nearer to understanding the true nature of this messy and worrying condition. However, both cases were highly memorable, and I'm keen to see more.

I have treated the condition as a response to stress from the presence of other house cats, even though there were no other signs of conflict and they were all happy to sleep together, groom each other, share food, etc. Sometimes the presence of friends can be stressful and competitive even though they are liked or encouraged. The vomiting might be likened to the vomiting of parent dogs or even certain sea gulls that regurgitate stomach contents to feed their young. Although this behavior is not reported between cats and kittens, it is about as close as I can guess to an explanation. Young puppies force their parents to vomit by worrying at their mouths until they get a meal, and perhaps the stress and unobserved worrying by even friendly cats causes vomiting in cases such as Ming's. Why it should be projectile, caused by rapid constriction of the stomach, is more likely to be a genetic feature and restricted to the morphology of the Siamese. Overly complicated conjecture perhaps, but in terms of attempting treatment very simple. We reduced all possible stress by isolating Ming and other sufferers during and after mealtimes and offered a secure, warm room for sleep away from the others. Meals were given in small, digestible doses and only up to early evening to ensure that stomachs were empty by nightfall. In one case sedative treatment was also prescribed for two weeks as an adjunct to help the cat cope with separation from its friends, as it was prone to be nervous when isolated at other times. In one of my cases this has solved the problem even to the point where the cat can now be fed later in the afternoon and early evening. Other cases of Hilary's and the other with which I am involved continue to defy all treatment, and we are currently exploring the prospect of keeping the cat more stimulated during the night by leaving it in a different room each night on a rotation of five rooms with the windows open, the lights on, and alarm clocks set to wake it up and stimulate it into moving to prevent prolonged relaxation. No results as yet; research is continuing into what I think is sometimes a very specific reaction to stress but is also the product of a combination of factors, physical and emotional, to which stress may contribute.

Trauma

Whereas stress is sometimes difficult to identify due to the large variation in individual responses to it and the apparent lack of any major or continuous behavior change in many cases, trauma is manifested by marked behavior responses. A physical shock from a wound is traumatic and causes the human victim to cry out, clutch the affected part, and show other instantly fearful or defensive reactions to ensure that the damage is restricted and the prospect of repetition is minimized. Emotional trauma resulting from pain or even an unsuccessful attack can be just as dramatic, especially in cats, and can even lead to physical collapse and death, particularly with sensitive individuals of high-strung breeds such as Siamese.

Interesting cases of "shell shock" have been reported in British cats that survived air raids during World War II. Similar to the reactions some people experienced after the blitz, the cats became withdrawn, unwilling to respond to their owners, and easily frightened by sudden movements or unusual or loud noise. As with other traumas one would expect that sound phobias developed afterward in many of these cases, even where gentle handling and encouragement improved the cat's abilities to communicate and more normal levels of activity returned.

A more bizarre and hitherto unique behavioral response to trauma was witnessed in cats that were subjected to a flood at an English coastal resort. They subsequently made repeated efforts to paw at and pounce on imaginary objects on the ground and in the air, in a manner similar to the imaginary fly-catching behavior of certain dogs. Cavalier King Charles Spaniels are particularly noted for this behavior, but without the excuse of trauma. Unfortunately no record of any treatment or the fate of the crazy coastal cats remains, so I must await my first case of this problem to see what might be done.

"Hello, is that Mr. Neville?" a crackly voice asked on the phone one morning. "I'm calling from the Middle East about my Siamese cats."

I'd be fibbing if I said that I didn't immediately have visions of being flown to some fabulously rich Eastern potentate's palace, there to administer at least a month's advice on treating a forlorn harem cat with a diamond collar. Alas it was not to be. But the problem about to be related was of enormous interest.

"Mo is my mother cat; she's five and spayed and lives with Ali, her neutered son, who is four. They've been together since Ali was born, and we brought them out here with us two years ago from the UK. They've always been totally devoted to each other; they sleep together, wash each other, and even share the same feeding bowl. Then last week my husband accidentally stepped on Ali's tail. Ali yelled out and immediately ran to hide behind a chair in our living room but was stopped by his mother, who attacked him savagely. My husband got badly scratched and bitten in separating them. We kept them apart for an hour or so to let things cool down a bit and then reintroduced them. Bang! Off they went again with Mo attacking her own son just as aggressively as before. So we left them overnight and tried again, but the same thing kept happening and has every time we've tried to put them together again over the last week. Just the sight of him seems to set her off. Ali is obviously very frightened now, although both of them are perfectly normal with us on their own."

We settled down to a long, expensive phone call to decide how to treat this very disturbing case of a trauma-induced breakdown in relations. It was ultimately resolved, but not quickly or easily and not without hiccups along the way when progress in getting the cats back together was undone instantly by a flashback and repetition of the attack by Mo.

Trauma, by definition, comes as a surprise. The body's startle reactions to being attacked are swiftly bypassed. There is no opportunity for the flush of adrenaline to be put to use in helping the victim escape as the damage is done before the victim is aware of what is happening. Instead adrenaline can cause the aftereffects of shock: lowering of body temperature, increased heart rate pumping blood faster to constricted capillaries and thus raising blood pressure, causing hair to erect and muscles to contract involuntarily, producing nervous shaking. Dramatic, yes, but thankfully rare as a cause of long-term behavior problems in cats. They either die at the time from the physical trauma of a road accident, for example, or soon after from the effects of physiological shock.

Cases that I have seen of long-term behavioral consequences of a single short traumatic incident usually concern relations between cats sharing a house, such as Ali and Mo, which can be totally broken down by something as fleeting as an accidental step on the tail by an unwitting owner. I am currently treating a similar case, also concerning two previously doting Siamese, whose relations were totally disrupted by a local rival entering the house via the cat flap. There was a single, short, serious fight with one of the resident cats. The other Siamese joined in, the visitor was swiftly defeated and chased out, but ever since this incident the first Siamese has attacked his friend on sight.

The reactions of the moment may be understandable in both cases. Ali, as a result of the shock and pain of being stepped on, cried out, and perhaps this was spontaneously frightening to Mo. She saw him as the cause of her own anxiety and therefore instantly deserving of an attack to force eviction and bring relief to her own fear. What is surprising, however, is the permanent effect on their relationship. Simply the sight of Ali was initiating the same pronounced level of reaction in Mo weeks after the traumatic incident without any apparent diminution of her aggression with time or frequent exposure.

Treatment involved total control of their exposure to each other on a very gradual basis. Baskets were used to house each cat individually to allow them to share the same room for very short periods. They were placed at opposite ends of the large living room; with the owners present to offer what reassurance they could, they were allowed to react in their predictable way. Mo was frustrated in her attempts to attack Ali, who in turn was prevented from running away. Confined, he remained calm while his mother screamed at

him and at least was unable to fuel her reactions by running.
Once Mo's initial reactions had subsided, the two were isolated from each other and the process was repeated just two
or three times per day for a week or so. They were left in the
same room for progressively longer periods and more frequently as the weeks went on. On some days Mo would not
react at all, and the cages could be brought closer together
for a short period. On other days her reactions were extremely violent, even to the point where she had to be
quelled with a water pistol, although there was a risk that this
could have alarmed her more. Gradually, however, the trend
was an improving one as mother and son learned to accept
each other again.

They were released in between sessions in separate
rooms and after about three weeks were allowed contact
across a wire mesh divide between their respective rooms.
Investigative sniffing followed, and things looked good, until
. . . the husband went through the mesh "door" and forgot to
secure it behind him. One scream and a huge scrap later
and we were back to square one. With husband suitably
admonished and kept out of future treatment plans, we began again, and this time with steady progress over several
weeks. When the great day came to put them together physically, it was actually done away from home in a friend's
house so that both cats would be a little subdued in an
unfamiliar environment. It seemed to work, and their mutual
acceptance persisted when the process was repeated back at
home.

The relationship between the two cats is still nothing
like their original closeness, even many months later. They
do not sleep together and are deliberately fed separately to
avoid any escalation of a warning grumble over a favorite or
remaining morsel. They are never left together unattended
as there is little doubt that Mo would inflict serious and
lasting damage on Ali if a fight went uninterrupted. She may
even kill him, so we take no chances. Steadily, however, the
two cats are getting closer again and from tolerating each
other's presence will now wash each other for a few seconds

before Ali, not sure of the wisdom of this, retreats slowly away.

There is no overcoming the effects of trauma by simply leaving such cats together to work it out, nor could any solution be achieved without much effort and patience on the part of the owners. The psychological effects of that one short accidental incident are far-reaching and probably permanent in many ways. Some owners would not have tried to repair the damage for too long and instead would have given one or the other cat away or tried to ensure that the two cats never met again; perhaps this is a sensible option in some cases. But at least it was interesting to see to what extent such a dramatic problem could be treated and how long it can take even with such willing owners trying so hard.

ᕲ 7 ᕬ

Aggression

The reactions of Ali and others to trauma have often been markedly aggressive, with subsequent attacks directed toward a single cat simply as a result of catching sight of him or her. By contrast the traumatized coastal cats exhibited undirected predatory aggression responses to their flooding trauma by pawing at the air and chasing imaginary prey. While these cases are rare and rather specific responses to obvious cues, they indicate the breadth of the range of behavior that we all too easily lump under a single heading of "aggression." When we start to consider "normal" aggressive responses, that range becomes even broader. Perhaps we are too quick to label certain types of behavior as aggressive because of the potentially dangerous consequences of attack.

We act defensively to avoid the aggression being directed toward us, and this can prevent us from distinguishing between different types of aggressive behavior.

When dealing with cats, it would be extremely unwise for me to treat all cases of aggression similarly, even though the actual behavioral response of each cat may be the same, with the use of claws and teeth. Each must be treated individually, looking closely at what triggers the response and its scale, duration, and direction, for treatment to have any hope at all. The general reaction of owners in the face of all types of aggression in animals is to become aggressive in return and try to subdue the cat or dog with greater force than being displayed, unless the animal is already beyond safe control or is likely to inflict injury on them. In nearly all such cases meeting aggression with excitement or aggression in return simply fuels the cat's or dog's aggression and has little or no reforming effect.

More often, and particularly with canine aggression, owners are attacked or bitten when they need not have been. While that may be their fault, the sad consequence is that the dog becomes aggressive more quickly when next challenged or more readily diverts other forms of aggression onto its owner. The end result is an upward spiral of aggression with the relationship in tatters, owners bitten, and dogs punished or even destroyed. Sadly the worst cases seem to involve owners who have received wrong advice from so-called dog trainers whose only answer to disobedience or defensive growls from dogs, frightened by overassertive training techniques, is to clobber the poor dog with an ever larger stick. However, most trainers worth their salt recognize that there are many different forms of aggression. They know that punishing frightened dogs that growl in an effort to deter the perceived threat is bound to frighten them more and instead use calming techniques.

Enough of dogs and my concerns over the standards of dog training. My point was simply to show how important it is to differentiate among types of aggression and never to treat it by applying aggression in return. This is even more important with cats, which, although far less likely to present problems of aggression than dogs because of that well-evolved ability to avoid conflict, are even better armed with a paw full of claws at each corner as well as teeth at the front. Potentially they are far more likely to inflict injury on us than most dogs if we get caught up with an angry one.

But just what is aggression in behavioral terms? Aggression is carrying out an act of hostility or causing injury to a rival, a prey item, or a threatening challenge. Indeed the term *aggression* is often used to describe only the first hostile act between two parties, though the victim may respond in like fashion and be described as defensive until he starts to win. Aggression is an integral part of every animal's behavioral repertoire, and it is expressed in a range of types and responses.

Body Language

Predatory aggression is vital for the survival of all wild hunters, as are other forms of aggressive reaction for defense when challenged. But aggression in a social or an interactive context between two adult cats or owner and cat is a different matter altogether. Indeed the repertoire of responses between two cats involved in a territory dispute may concern not only obvious displays of aggression in actual fights but also a number of fearful responses designed to ameliorate the aggression in the competitor. The interplay between the two rivals is often enthralling to watch (provided actual fighting doesn't occur) because of the dramatic tension and the range of responses in both cats. Body movements and postures, facial expressions, and vocalization are all used to communicate one cat's proposals or emotional state to another. The other will respond with equalizing reactions if standing firm, raise the stakes with more assertive gestures, or show retiring responses in deference to the other cat's assertive gestures. Such reactions are probably best described as agonistic rather than aggressive displays as they are actually designed to prevent violence. The bluff of staring an opponent in the eye, settling in to a fixed tense position as if ready to strike, erecting the hair, arching the back, and attempting to make oneself look larger and more imposing through sideways presentation, coupled with distracting tail swishes and low, threatening growls or whines from a crouched position, are statements of intent designed to convince the opposition to withdraw and so avoid actual physical interaction.

The squaring off of strutting and staring that often lasts only half a minute or so between agonistic dogs can last half an hour or more between cats. Often two rivals will settle into prolonged staring matches from apparently comfortable positions, crouched on the ground, and vary the sequence and combination of their agonistic gestures and sounds until one edges away—almost, it seems, as much from boredom as from discretion. The reason it takes so long for such interplay to prevent an aggressive attack from either cat, compared to the relatively short tolerance of high-ranking or territorial dogs, is probably because of a lack of social cohesion and development of formal body language that keeps a dog pack together as a hunting unit.

Cats, being solo predators, have not evolved a social system of dominance/submission interaction for any purpose other than occasional territorial or resource defense. Instead most cats use that discretionary avoidance of conflict to ensure survival by keeping out of harm's way. Indeed some authorities on cat behavior do not recognize submission in the cat because it retreats when outmaneuvered by a rival rather than rolling on its back like a submissive dog. I think that gestures of submission are given off vocally, and through eye movement and body language prior to this, but we are as yet unable to translate the subtle combination or sequence of feline communications when they are all at work. Some evidence for this is that a slowly retreating cat is usually allowed to withdraw by the "victor" without being attacked, yet cats that try to run away, a very submissive gesture practiced by many losers in the animal kingdom, invariably stimulate the "victor" to chase and attack to enforce the message. Or is it just that movement easily elicits aggression in cats and the roused, territorially agonistic cat simply switches into predatory chase aggression given a moving target? I doubt it, but as yet no one seems to know for sure.

Aggression in terms of actual fighting between cats is rare because even a very confident, dominant individual puts him- or herself at risk from a defending cat's claws and teeth while attacking, so it is safer to resolve disputes with agonistic displays. Fights are more common between males, especially unneutered ones. This form of aggression is predetermined by genetic influences as a result of the early masculinization of the fetus brain with testosterone hormone in the early definition of the kitten's sex. Later production of testosterone in the adult male cat causes the development of secondary sexual behavior patterns, one of which is fighting with other males to establish a territory containing survival resources and breeding females. Castration removes much of the secondary hormonal motivation for this behavior by removing the source of testosterone, as well causing the size of the cat's territory, and therefore the number of border disputes, to decrease. Most of us have our tomcats castrated at adolescence and don't own serious fighters (or smelly indoor sprayers, which is another largely testosterone-inspired pattern). Indeed I have never had an unneutered tom referred to me for aggression problems, and those that I have seen are always stud pedigree cats kept safely in a pen and denied the opportunity to meet or fight other toms. Those owners who can live with an unneutered free-ranging tom will have to accept that he comes home less often and may be frequently bloodied, bitten, and scarred from fighting. Abscesses often develop from deeper bites and scratches, and owners must simply be prepared to patch their tough guy up for the next encounter. This always seems rough in high-cat-density areas in our towns, particularly as the cat gets older and less able to fight. Castration reliably reduces the motivation to be aggressive toward other males. Fortunately over 80 percent of pet cats in the U.K. and the U.S. are neutered, and so few toms are forced to suffer as a result of their hormones in the overcrowded territories of suburban cat land.

Classifying Aggression

The classification of aggression problems in cats falls neatly into two distinct types—"normal," or expected under certain conditions, and "abnormal," or excessive reactions. The concept of what constitutes any form of problem behavior is largely a matter of opinion for each owner, but with aggressive cats the opinion is more readily determined by the risk of injury, and so tolerance levels are often lower. Perhaps for this reason I seem to encounter a greater number of cats presenting various forms of aggression than most cat owners would believe. Most cases do not concern normal or expected aggression, as most owners are aware of their cat's armory and predatory status, but nonetheless a few cases of the aggressive behavior that has been lost to most cats in the process of domestication and ordinarily is seen only in unneutered toms do filter through.

As with many aggression problems it is not always the violence of the problem that causes the upset for the owners. More often it is the contrast with the cat's normally placid, friendly countenance that comes as such a shock, and it is the inability to intervene with the usual friendly, affectionate overtures that causes owners to feel detached from their cat.

Predatory Aggression

Dear Mr. Neville,

My cat Arthur is an absolute murderer. He brings home at least one mouse, vole, or bird every day and has even managed fully grown rabbits and pheasants. I know that cats are predators, but is there anything I can do to reduce the level of his hunting apart from keeping him in all the time?

Yours sincerely,

Maureen Beany

Predatory aggression is probably the most accepted form of aggression in our cats. None of us like it, most of us feel guilty about it, yet most of us realize there is precious little we can do to prevent it. Some cats are better hunters than others, and a few are either not interested or just very poor at it. Some catch their prey, play with it, then lose interest and abandon it once dead—these are presumed to be non-hunters by their owners. Most, however, bring at least an occasional trophy home to their safe feeding lair, and some go as far as consuming most of their spoils. Hard rodent heads and a gallbladder left on the floor are all we usually find, though again some give up once indoors and abandon their victim in favor of tastier, passive cat food in a bowl. When hunting, handling, and dispatching their prey, cats exhibit none of the prolonged staring, puffed body language, or vocalization of the agonistic conflicts between cats or when fighting. Rather it is a quiet, efficient affair with stealth and speed the key to success.

Clearly cats have been officially appreciated in the very highest of circles for their hunting ability, though not all owners really like to face up to their pets' more bloody activities. Generally our attitude is to put such predatory behavior down to the cat's basic instincts and to wash our hands of the carnage unless the cat is like Arthur and brings an enormous volume of carcasses home. Others may bring them home alive and, having lost interest, release them to scurry or flutter around in the house or even take what seems to us to be a sadistic pleasure in torturing a half-dead victim on the lawn. Then we interrupt and, having grabbed the cat or chased him off, feel like St. George as the small furred or feathered victim escapes to live another day.

An American biologist who decided to investigate the environmental impact of pet cats living in a small village in Bedfordshire, England, came up with some alarming results. Over the course of one year 70 free-ranging pet cats in the village of Felmersham brought home no fewer than 1,090 prey items. Owners who dutifully complied with Professor

Robert May's request to bag up each tiny victim for identification were informed at the end of the study that their darling cats had accounted for 535 mammals, 297 birds, and 258 other unidentified small and furry organisms (UsFOs!). For the statistically minded this gives an average figure of 15.6 victims per cat, though this includes only prey actually brought home. The real figure, including prey items killed and abandoned away from home or totally consumed, will certainly be higher. One American study estimates that cats bring only half their victims home, but if we extrapolate the figures as presented to cover the U.K.'s total feline population, we find that its pet cat population of 6,800,000 kills 95,288,400 mammals and birds annually, allowing for 10 percent of cats that live permanently indoors and have no opportunity to hunt. Similar data were produced by British scientists headed by Professor John Lawton of the University of London. Studying the predatory habits of 77 cats in a village in Bedfordshire, they found more than 1,100 prey items were brought home, 16 percent of which comprised sparrows. They considered from this that predation by cats accounts for between a third and a half of sparrow deaths and that cats kill at least 20 million birds a year. This is interesting when we look at how outraged we are at the bird trapping and shooting by European Community fellow members around the Mediterranean.

The ecological impact of this pet predator is so startling because it fills a niche in Britain's rural and urban ecosystems that is otherwise unoccupied. The only natural small predators Britain has are stoats, polecats, and pine-martens, all far less common than even fifty years ago, although foxes and some owls are making successful comebacks after learning to adapt to the urbanization of the countryside. The U.K.'s natural wildcat has been in serious decline for many years and is now probably lucky to catch 15.6 victims per day in its last areas in Scotland.

Bolstered by all that lovely food we give it, the pet cat is enjoying exercising its unforsaken predatory prowess to the full in the perfect hunting reserve of Great Britain, and as we acquire more cats as pets, we stack the odds against our small mammal and bird population. One last thought. In 1988 a little more than 2.7 million rats and mice were killed for experimental purposes. While we rightly feel outraged and campaign for the adoption of alternative testing techniques in laboratories and cruelty-free cosmetics, our cats are outside killing 35 times as many, and not even for essential food.

So, Maureen, I'm afraid that we are rather up against that uncompromising 13 million years of evolution since the first cat walked the earth in trying to reform Arthur's murderous habits. Even 5,000 years of domestication, and selection for friendliness and beauty in the past 100 or so, has done little to alter his instincts. All we can try to do without frustrating all his outdoor activities by keeping him permanently indoors, which should never be regarded as an option with a cat used to the outdoor life, is to put him at as great a disadvantage as possible. A bell worn on his collar may give some targets a little more time to escape, and restricting the cat's opportunities to hunt by keeping him in at times when rodent populations are most active, at dawn and dusk, may also reduce the carnage. Temporary boarding for a few weeks or so is not unheard of during the nesting season to protect young birds.

Another idea to calm the hunting instincts is to feed the cat a good full meal of favorite irresistible food before he is allowed out so as to slow him down a little. This I call the "Alcatraz tactic" after the famous island prison's policy of feeding inmates large volumes of fattening food to keep them unfit for escape and unlikely to be able to swim across San Francisco Bay to freedom if they did.

Feeding the predatory cat fresh, gristly meat still attached to the bone forces him to spend much more time handling and ingesting his meal, which may also reduce the desire to do it again with live prey. But perhaps Maureen will sleep more easily if she prevents herself from having to witness Arthur's "crimes" by closing the cat flap and denying him direct access to the safety of his feeding lair. Couple this with shooing him away from the back door with a hiss and a small cup of water when he tries to bring a victim home, and this may at least persuade him that the kitchen isn't such a safe place to bring his spoils. Perhaps better still is to obtain a kitten from a nonhunting mother and keep it indoors and unexposed to rodents and birds until about one year of age. This tactic is reported to produce cats with an undeveloped interest in birds at least and undeveloped skills in tackling rodents. However, the lack of opportunity to develop in other respects may also produce a rather dull cat as a pet, and if it catches up and becomes more interesting through new opportunities after it is allowed out I suspect that in time it will learn to hunt, even if not as effectively. Movement of small beasts and hopping birds will still arouse chase instincts. I have treated several older cats (for other problems) that, when allowed out late in life following a move out of the city to the country, have quickly proved themselves to be very efficient hunters.

> Dear Mr. Neville,
>
> My cat is called Mouse, which is very appropriate because she is very good at hunting. I know it's natural, but can you tell me why she has to play with the poor little things and torture them to death? Is there any way of training her to kill things more quickly and humanely if she must hunt?
>
> Yours sincerely,
>
> Kay Brown

The usual hunting strategy of a cat is well known to us all. Attracted by the scuffling noise of a mouse in the undergrowth or the movement of a vole scurrying past or a bird hopping along the ground, the cat freezes momentarily and stealthily creeps toward it, crouching low to avoid detection. The eyes fix on the target, and with a quick-aiming, muscle-activating wiggle of the rear end the cat achieves a lightning-strike capture for about one in three small rodents and slightly fewer birds, who have the advantage over the cat of being able to fly. The prey is grabbed with hooked claws and quickly bitten, but what follows is largely a function of the cat's experience. Mother cats that are proficient hunters raise proficient kittens, able to recognize prey items at about a month old after she has brought home a few items to smell, and then taste, in the process of weaning. She teaches her kittens the art of the swift nape bite to sever the spinal cord while keeping the rodent's teeth facing away from the cat and therefore unable to bite back. The nape bite is designed to dispatch the victim quickly, ready for eating. The ability is taught and developed by offering a wiggling tail to encourage chase and pounce by the kittens, and then the half-dead prey is brought back to the nest for them to learn to handle and dispatch. A few misplaced bites from the kitten earn a return bite from the victim, so through guided trial and error it soon learns where to bite to effect the safest and quickest method of immobilizing the rodent.

Inexperienced mothers may not offer a wiggling tail very often, and inexperienced or poor hunters may never bring enough half-dead prey on which the kittens can practice. The young cat therefore never progresses beyond an instinctive interest in movement. Such interest may even cause a cat to abandon a dead mouse to pursue a live one, but, having caught it, its interest may be maintained only as long as the mouse struggles and tries to run away. As the mouse grows weak from being pounced on repeatedly, the cat may even try to "activate" it to maintain the cat's interest by throwing it in the air or allowing it to get away a little before chasing again with the apparent aim of encouraging the mouse at least to try to escape. Once the mouse is dead, however, many cats lose interest and do not continue to play or go on to consume their victims.

In a world without free food in bowls such inept hunters would be ill equipped to survive, and only the proficient hunters would live long enough to produce kittens to which they can pass their skills. Perhaps in the last few decades we have been finally watering down the predatory side of the cat

by breeding from mother cats with poorly developed hunting
skills. But it will not be a short process. Many cats make up
for early deprivation, and we deny most the opportunity to
pass on their hunting incompetence to successive generations by neutering them.

Perhaps the best hope for producing a nonpredatory cat
lies with the pampered pedigrees, kept fertile for breeding in
sheltered houses or pens without the opportunity to develop
prey-handling skills. But somehow I'm sure that even their
kittens will retain the potential for interest in chasing mice,
rats, voles, shrews, birds, rabbits, and pheasants given the
opportunity. As a result I suspect that an even higher proportion would, like Mouse, "torture" their prey, never learning
the swift dispatch techniques of the nape bite, and so continue to disgust their owners given the chance.

Maternal Aggression

Dear Mr. Neville,

*When Lucy, my Birman queen, produced a litter
in May, she became very aggressive toward all of
us and wouldn't let us near the kittens. While I
understand that it is probably natural for her to
defend her kittens, it seemed that she didn't recognize us and even attacked us when we were
just walking by, not even trying to take a look.
Friends who also breed cats said they'd never
seen such a change in a normally docile, friendly
cat, nor such a defensive mother. Do you think
she was upset by this? Should we breed from her
again?*

Yours faithfully,

Susan Swift

Maternal aggression is another of those acceptable forms of
aggression with which we can readily identify and so toler-

ate, even though the scale of the attack on us in a different context may be totally unacceptable. The majority of female cats don't defend their kittens to such a degree as Lucy, but when they do, it can be one of the most pronounced types of aggression we are likely to encounter. While most are happy to let us into that most special of areas, the nest, others that feel challenged by our approach have to mute their own flight responses to stay to defend their genetic investment for the future. The pent-up emotion can then suddenly explode with a ferocious attack, without warning, either at us or at other cats that are ordinarily tolerated or even accepted as friends.

Dr. Benjamin Hart, from California, one of the founders of animal behavior therapy, suggests that such aggression may be released by the fall in the level of progesterone hormone at the time of birth. During pregnancy this hormone has been produced in large quantities to help the mother's body cope with the physiological changes involved. Progesterone also acts as a calming influence on the areas controlling emotions in the brain, and so with this capping effect lost the mother may bounce back to become highly reactive and easily provoked into emotional reactions and defensive behaviors. In humans the loss of progesterone levels is implicated in some mothers who suffer from postnatal depression, so perhaps there is a link, and perhaps in cats some types of maternal aggression might be treated in the short term by the administration of artificial progesterone, known as *progestins*, from the veterinarian. Otherwise it's a case of providing as secure a nesting area as possible for cats like Lucy and not attempting to handle kittens until some days after birth, and then only when the mother is away, until she is more relaxed about our passing and approaches. Interestingly, however, cats like Lucy do not always respond in the same way with each litter and may be perfectly tolerant of people approaching their next kittens. Far from being overprotective, other female cats with poor mothering ability may neglect or abandon their kittens entirely, and then we are forced to take on the whole role of bringing them up. Again we are propping up incompetence in cats, for in the wild such poor mothers would never reproduce successfully.

Dear Mr. Neville,

*When one of my eight cats, Charlie, had a litter,
she ate them all after four days. Why should she
do such a horrid thing, and should we have her
spayed now so that it can't happen again?*

Yours faithfully,

Pat Bream

Cannibalism is recorded in many species, including mice, rabbits, dogs, and, more rarely, cats. Although abhorrent to us, it can have perfectly reasonable explanations. One may be a failure of the hormone system to inhibit killing of prey at the time of birth and shortly afterward. Another, in feral colonies in particular, may be that the drive for self-preservation overcomes maternal instincts in malnourished mothers. Litters resulting from the second pregnancy in a year are apparently more prone to being cannibalized, as are those that are sickly and perhaps not worth the maternal investment to raise them, according to Ben Hart. Larger litters are more likely to be cannibalized, but more usually cannibalism occurs as a result of the failure of the mother to find a secure nest area in which to nurse and raise her litter. Stress caused by overcrowding lies at the root of many cases of cannibalism, and Charlie probably fits this bill. Next time she should be given a safe, quiet kittening area away from all other cats at least two weeks before giving birth, and perhaps she will then take good care of her offspring.

Cannibalism of kittens by tomcats is often cited but is rarely experienced in pet cats. In prides of lions, however, new dominant males taking over a group of females will systematically kill all cubs sired by the previous male in a ruthless purge. This was recently filmed for the first time in Africa and shown at peak viewing time to the British public to much debate about the acceptability of such violence at such a time, even if it was a true-nature program. The aim of the infanticide is to ensure that cubs with no genetic input from the new male do not drain his reserves in helping to

fend for them and feed them. Instead the mothers of those
cubs came into season again within twenty-four hours of the
massacre and were mated by the new male. They subsequently gave birth to his cubs, which, deserving of his protection, ensured that his success was passed on in the form of
his genetic complement, not his rival's. Such behavior is also
reported to occur in the domestic cat when living in social
feral groups but rarely happens in our homes as in many
cases they are protected from the roaming and perhaps long-
gone father or, if resident, he is safe around his own kittens
anyway. Nonetheless a most important way of preventing
cannibalism is to provide a secure resting area, safe from
other house cats, especially males. This is crucial during the
first few days after birth and can be relaxed with friendly cats
later. Introductions should be supervised, and if the mother
is intolerant permanent protection should be afforded until
the kittens are weaned.

> *Dear Mr. Neville,*
>
> *When our cat Belle had her kittens, we decided
> immediately that we would keep one particularly
> pretty daughter as company for her after all the
> others had been homed. Although Belle was an
> exemplary mother, she now won't have anything
> to do with Tina, who is now six months old, and
> even spits at her and tries to chase her away if
> she comes into the same room. Why should Belle
> be like this to her own daughter? Is there any-
> thing we can do to make them like each other, or
> should we look for a new home for Tina?*
>
> *Yours sincerely,*
>
> *Caroline and Colin Barratt*

Poor Caroline and Colin. It seems that Belle is one of those
highly territorial females who cannot tolerate competition
for food, shelter, and, perhaps most important of all, kittening area. While some females will accept any number of

cats, related and unrelated, and share even in the nursing and upbringing of other kittens, a few are intolerant even of the presence of their own offspring once they are no longer kittens. There are none of the human family values of protecting children into adulthood with cats. While the kittens are dependent on their mothers for food prior to weaning and for protection and development for a few months afterward, cats such as Belle make as good a mother as the next cat. But once their own offspring start to mature and become nutritionally and behaviorally independent, the mother will try to edge them out of her territory to establish their own and ensure she keeps all those resources that she will need to keep herself and any subsequent litters. While this perhaps indicates forward planning, the results are often spectacular, with the levels of aggression demonstrated very marked. Occasionally a youngster may learn to fight back and establish a right to share in those resources again and even in time to enjoy quite a close relationship with its mother. However, the chances of this are probably about equal to those of successfully introducing an entirely strange cat, and the cat will receive no special favors for being related to its mother.

In feral colonies the expunged daughters may establish territories next to that of their mother, and subsequent encounters at the borders may be reasonably amicable. My own observations of feral cats and experience with cases referred would seem to indicate that male offspring are more likely to be tolerated for longer by territorial mothers than females, but there is no other evidence to support this. When female offspring become sexually competitive with their mothers, the fur may fly, as then the resources for raising kittens are seriously under challenge. As males will not present this form of competition, they may be tolerated longer, though ultimately they too are driven away as a natural mechanism to prevent inbreeding.

But for Belle and Tina the prospects of harmony are not good, and integration is probably impossible unless both cats are spayed. But even then tolerance of each other may be all that can be achieved, and the cats will never be seen to play or sleep together or groom each other. In fact it is possible that they will never be relaxed when together, and therefore the house will always be in a state of tension to the point where it is better from everyone's point of view to comply with the natural process and find Tina a nice new home. It's a pity, especially when so many cats do get on so well together and bond very closely and permanently to related individuals.

Food Guarding

Dear Mr. Neville,

When I feed my tabby cat Campbell, he crouches low over his bowl and growls fiercely. He eats in short, quick gulps, pausing after picking up a lump of meat to look around. If I am close, he growls again, and if I try to stroke him he drops the food, hisses at me, and has even lashed out at me with his paw. Of course I now leave him on his own to eat his meals, but is this common in cats? I had a dog once that used to guard bones, but I managed to train him out of it. Could this be done with Campbell?

Yours sincerely,

Simon Blake

Food guarding is more a feature of canine behavior than feline. Bones are the ultimate tasty package and, being none too easy to consume, invoke a guarding response to ensure that ownership is retained long enough to get at the goodies. Dogs are also more likely to guard high-quality canned or fresh diets than less palatable dry or cereal-based canned diets, and diet manipulation is a key feature of treatment. A few cats also guard their food, especially when in direct

competition with other cats at the bowl. Ears may be flattened, growls emitted, and agonistic body postures adopted,
but by and large the subordinate individual bows out without
coming to blows. Standing low over the bowl and growling at
the owner who has only an instant before gone to the trouble
of opening the can and dishing it out to a hungry cat may
seem ungrateful but is a natural and easily explained agonistic reaction designed to preserve resources. Feeding more
frequent, small volumes of food that are more quickly consumed may lower the intensity of the reaction, as will providing continuously available dry diets, which involve not only
less concentration of feeding opportunity but, being less
aromatic and attractive, are less likely to be worth guarding
anyway. As with dogs, however, all conflict at the food bowl
must be avoided, so Campbell and others like him are best
fed on their own and left undisturbed.

Pain-Induced Aggression

Dear Mr. Neville,

*I have tried to groom Snowflake, my Persian cat,
every day since I bought him at thirteen weeks of
age. Now he is a year old and still hates being
groomed. He wiggles and curses and then
scratches and bites me. I've tried to be gentle and
offer him tidbits, but nothing seems to calm him.
After a while he is so wound up that he grabs my
hand and arm and really bites hard. If I hold on,
he bites harder; if I let go, he runs away and
won't let me near him. Is there an easier way? I
really do have to groom him; otherwise his fur
gets tangled and matted.*

Yours scratched,

Helen Ford

Difficulties with grooming or trying to treat an injured cat
are the main occasions when we are likely to encounter

pain-induced aggression. It is understandable that injured cats will try to avoid further threat and make us keep our distance. Pain elicits an aggressive reaction, as we know if we accidentally step on the cat's tail or inadvisedly smack a cat for a misdemeanor. As often as not we get an immediate swipe in return, and our efforts to correct the cat's behavior fail because punishment has no therapeutic role with the adult cat. If the shock is too great, the cat may also be traumatized for some time and avoid our efforts to make up through a prolonged fearful reaction. Traumatic painful episodes can lead to low-threshold, specifically targeted aggressive reactions, as discussed in the "Trauma" section of Chapter 6. Prolonged exposure to a less startling painful stimulus can also produce a progressively decreasing reaction as the cat habituates to the pain and realizes perhaps that any counteraggression to it brings no relief. Early play fights between kittens help them control their responses to pain and associate their own biting of a littermate with a painful counterreaction of aggression in return. Kittens brought up in solitude do not acquire this awareness of the influence of others and consequently are often spiteful as pets and easily roused to aggressive attack when handled by us.

Grooming, however, is a different matter, especially when we've tried to be gentle in what, with long-haired cats, is an essential daily aspect of care. While some enjoy the attention and handling, others, like Snowflake, never accept it readily and, like the solo or hand-raised kitten, are always aggressive in defense. The cause almost certainly lies in the cat's upbringing. Snowflake's breeders probably didn't acclimatize him properly to being groomed from the first day they could handle him, when those feedback mechanisms were open for manipulation. The fractious young kitten can be disciplined with a firm no or a mild countertap if he lashes out, in the same way as his play aggression releases a corrective, aggressive response from his littermate. Tolerance develops, and this can be rewarded with gentle handling, and later on, with tidbits and treats. If this process is delayed, the opportunity is missed, and forevermore the cat will perceive the brush and comb as a major threat, always

deserving of an aggressive response. As a result many owners have to resort to sedating the cat for the purpose of
grooming or, worse, leave it as long as possible to get dirty
and matted and then have the cat clipped or shaved under
anesthetic by the vet.

Though the days of getting Snowflake to enjoy being
groomed may be long past, there are nonetheless alternatives to sedation in making grooming safer for the owner.
First the cat should be placed on a perfectly steady table at a
comfortable height for the owner. On the table should be
placed a covering of thick carpet. The cat should be pulled
gently backward as instinctively he will try to hang on by
extending his claws into the carpet. It may be a two-person
job, but the initial grooming should then be very light along
the back and not directed at sensitive areas such as the
abdomen until much later. Mats should also be avoided in
the first few "mock" grooming sessions. Gradually the brush-
ing can become more penetrative, always accompanied by
gentle voices and the prospect of treats, though it has to be
said that few cats will be very interested in them until more
tolerant of the handling. As with cases of aggression in
kittens and young cats, where owners simply must persist
with their approach to grooming or handling an adult, a
steady increase in frequency and intensity of handling is
most likely to subdue the cat's aggressive responses compared with occasional sudden, high-pressure efforts. But
when this tactic fails or is still providing the owner with
scratched arms, a hood over the cat's head will have a calming influence and at least enable the job to be done, even if
the cat is not actually learning to accept the handling.
Several types of hoods are currently on the market. The
typical reaction of the hooded cat is to crouch low and move
only very slowly, if at all. Gentle grooming can then be
introduced safely, as can other forms of invasive handling
such as the administration of ear drops, to which cats often
respond with aggression if able to see. Similar responses to
darkness are also utilized by hooding birds of prey and to
calm horses that are reluctant to enter the stalls at the

racecourse. Ideally, however, all kittens should be handled frequently and, especially if long-haired, groomed as early as possible and well before six weeks old.

Stroking and Aggression

The strange nature of the threshold between tolerance or even enjoyment of handling and aggression in cats is well illustrated by what the Americans call the "petting and biting syndrome" or "petting-induced syndrome." Most affectionate pet cats can be encouraged to exhibit this type of aggression quite readily.

The cat, which in America at least is more likely to be male than female, apparently enjoys a cuddle, settling comfortably in its owner's arms and purring loudly. Stroking the head and back could go on forever, but touch the abdomen or tickle the hind legs and nearly all cats will suddenly grasp your hand or arm, scratch and bite, and sometimes kick with a repeated movement of the back feet. Others respond in a similar fashion to far less physical interference, sometimes for simply trying to pick them up. Most cats will at least tolerate being picked up and enjoy being cuddled for a short period, but from lying in your arms relaxed and happy one minute the cat may become instantly violent, looking almost as confused as we are. As we curse or yell at the traitorous little so-and-so, it has already jumped down from our arms or lap and may stand a few feet away, with ears back and not sure of its next move. Often its move will be to sit and groom enthusiastically, which is sometimes a diversionary self-directed behavior that cats enact when confused or slightly nervous and sometimes an effort to wash or spread the smell the cat has collected from your hands more evenly over its body.

For cats that are used to our nursing attentions the threshold of this reaction may be very high and only rarely encountered. For others, perhaps unused to handling or handled less frequently as kittens, the reaction appears within a few moments of being petted and without our having to touch those more sensitive areas. It seems that the threshold of the reaction is reached when the cat ceases to feel comforted by our mothering-style affection and suddenly feels trapped and vulnerable so close to us while in such a relaxed state. From having allowed itself to revert to its kittenhood in that maternal-kitten relationship we encourage, it grows up in an instant and as an adult solo predator decides that it needs to repel what has become a mild threat and put some distance between us. Small wonder it looks a bit confused as it jumps down.

Theoretically the potential for this reaction lies in every pet cat. Treatment simply involves trying to raise the threshold of the reaction by never reaching it. Cats with low thresholds of reaction should be fussed over in very frequent, very short encounters, initially not even involving picking them up. If they will accept this, they should be encouraged to lie sphinxlike on one's lap and be stroked on the head and along the back. Escape is easier from this position and does not involve having to beat us away before the cat can get down. Gradually the length of such contact can be increased so that the cat learns to feel more comfortable in physical contact with us, but only much later should attempts be made to invert it to be cuddled. Those sensitive areas should never be touched and the cat never restrained if it decides to depart, as both of these actions will lower the threshold of "petting and biting syndrome."

Learned Aggression

Dear Mr. Neville,

My little female cat Rose is usually very quiet and affectionate, but whenever she sees my neighbor's dog, or any other dog for that matter, she immediately attacks it. I thought cats were traditionally afraid of dogs, but where I live it's the other way around! Why should Rose be like this, espe-

cially to the dog next door, who really only wants to be nice to everyone?

Yours faithfully,

Sarah Plumb

Dogs and cats! From Tom's encounters with Spike in "Tom and Jerry" to the Peke-faced Persian, which is a bizarre attempt on our part to select a cat that resembles a dog, to the not infrequent but certainly impossible claims that someone has managed to interbreed the two species, we wonder about the relations between our two favorite pets. In cartoon town, "Dogs hate cats," as Spike says when instructing his pup in one of those marvelous bits, yet many of us who keep both know that they can be the best of friends. When problems arise, it is usually from the canine end, as many dogs will chase cats or anything else that looks like it might run.

Faced with a chasing dog, most cats try to avoid the conflict and run for cover as fast as they can, preferably over and behind a fence or up a tree where, once safe, they glower, highly displeased. A few learn that running away only heightens the dog's predatory excitement and stand their ground with ears back, body arched, threatening to strike. There is nothing quite so funny, as the makers of all those cartoons are aware, as when the charging dog suddenly realizes that this is no ordinary cowardly cat and that the chase should stop. The problem is stopping in time to avoid upsetting the cat, before the roles are instantly reversed. The dog, his pride in tatters and glancing around to ensure that no one witnessed its failure, withdraws and trots back to its owner while the cat settles to whatever it was doing before the dog appeared.

Unless, that is, the cat is like Rose. Perhaps a dog did just nip her one day, or perhaps she was sufficiently frightened by a swiftly advancing menacing mutt or cornered and forced to beat off the challenge by unleashing a defensive swipe with a fully clawed left hook. The success of her aggressive defense saved the day and became incorporated into her behavior repertoire as a learned pattern of aggression. But now, rather than waiting for the threat to manifest from an apparently relaxed, nonthreatening dog, she wades in first to force a retreat before it can attack—a preemptive strike to preclude any possibility of being caught at a disadvantage. Now the shoe is on the other paw, and it is the dog that must learn the techniques of conflict avoidance. If necessary, treatment may be possible in acclimatizing Rose and her ilk to regularly encountered individual dogs such as the one next door by penning her nearby and allowing the dog to approach and sniff without being chased off. This is similar treatment to that of desensitizing cats to visitors described in Chapter 4. But while this may be successful with the one dog, it may not be so for all the others who dare come within range of cats like Rose.

Dog chasing is just one example of learned aggression in cats. Another is the ever-quickening response of aggression to having your tail pulled repeatedly. This can be rather dangerous with young children about. Although the cat usually learns to find high sleeping places after a couple of assaults from a crawling or toddling child, the child may still pursue the cat as a moving item of interest every time it appears in the room. It is natural for the child to show that kind of interest, but it can be a frightening phase for the cat. The child wants to make contact but has little control over the strength of his or her patting or grabbing. The cat usually learns quickly that an aggressive swipe or bite deters such unwanted attention as the child screams in pain and stops advancing.

As with the chasing dog, the cat may soon learn to get in first and advance on the child to preclude all possibility of heavy handling. The cat perceives this as being a better option than trying to avoid the conflict in the usual way by running off. In some cases this type of learned aggression can be very hazardous indeed for a while but may gradually dissipate as the child grows up and gains more control of his or her movements. The cat often comes to accept normal petting from the child later in the normal way, but a few remain aggressive at the sight of children, and not just those in the cat's own family. Visiting children may be under just as much threat, and so in these cases the cat will need to be separated from family life to prevent injury.

Of course no cat or dog should ever be left alone unattended with a child for both their sakes, but while such aggressive cats can be managed at home with a little care, it is probably safer in most instances to place the cat in a child-free house. Physical punishment of the cat by the child's parents only serves to heighten and spread the cat's aggressive reaction and will have no reformative or reconditioning benefits at all. Intervention from a dominant member of the group to induce a submissive response does not work with cats in the same way as a well-timed vocal or startling intervention may do with a dog in a similar aggressive frame of mind. Aggression from us causes the cat to be fearful and lash out to protect itself even more.

Dear Mr. Neville,

When we first got our cat, she was only a little scrap from the local animal shelter but soon fit in well with our house and went on to carve out a territory in the local gardens. She bonded extremely closely to me and initially accepted visitors and my boyfriend very well. She was always quite rough when playing and seemed to be more ready to scratch and bite my boyfriend than anyone else. Since then I've changed my boyfriend, and he has moved in. He adores cats, but Chloe hates him despite every effort on his part to be nice. She physically attacks him whenever he walks in, and if I hold her she glares at him and growls. Now she doesn't like visitors either and attacks people she used to like before, even though she is still very affectionate to me, to the point of following me around the house and trying to sleep by my head at night. My problem is how to reconcile Chloe with my boyfriend and quickly, because we plan to marry in a few months' time!

Yours sincerely,

Christine Allan

We have dealt with such specific dislikes earlier, with cats whose reactions are nervous and withdrawn. Specific aggressive reactions are less common, though every cat will have its likes and dislikes of different family members and visitors. To some extent Chloe's aggressive threshold may be low because of encouraged rough play when young and success of the ultimate kind in having disposed (perhaps in her mind) of the previous boyfriend as well as learned success at repelling visitors with an aggressive display. The aims of treatment are essentially to acclimatize Chloe to the boyfriend gradually, perhaps with her protected initially in a pen or on a harness so that she can grow used to his right of access and presence. Christine, too, has a role to play in detaching herself from Chloe and relinquishing the roles of feeder and affection provider when her boyfriend is with the cat. Initially frequent short meals should be the only advances made by boyfriend to cat, and he should not attempt close handling until much later. Low and decreasing doses of tranquilizers of the type more commonly used to control car sickness in dogs seem particularly helpful in such cases, provided the cat receives frequent short exposure to the boyfriend. So if marriage is in the offing, this should be an aid to progress, and the boyfriend need not be traded in for one that Chloe finds more acceptable!

Idiopathic Aggression

Dear Mr. Neville,

Our Burmese cat, Gong, is a pleasant, friendly cat for most of the time. As we live in an eighth-

floor apartment, he never goes outside, but he
enjoys a good life with us and receives much
attention and petting. We give him only the best
food and lots of toys to play with, but approxi-
mately once a week he attacks either my hus-
band or me as we walk into the room where he
happens to be lying or sitting. The attacks are
quite fierce and aimed at our legs or arms. Gong
seems manic during the attacks, which last only
a few moments, and he takes some time to calm
down afterward. We live in fear of him some-
times, but at other times he is the most respon-
sive cat there could be. Is there a way of making
him safe?

Yours sincerely,

Clare and Steven Hall

Nearly all the cases of ostensibly motiveless aggression that
I have treated concern Burmese cats that live permanently
indoors. The majority of the others are moggies but, again,
are indoor cats. While it is unwise to rule out the possibilities
of fear-induced aggression in such cases, perhaps caused by
the owners' entry into the cat's sleeping area and startling
him, most do not fall into this category. The aggression is
what scientists call *idiopathic*, which means "no known
cause." In short we haven't got much of a clue as to why it
happens, but we couch it in impressive-sounding language to
cover up.

There are medical conditions that may cause a cat to
react in this way, such as hypothyroidism, and in this case
thyroid hormone replacement therapy can resolve the problem, but this is obviously a job for the veterinarian. Other
causes include forms of epilepsy, which can be treated successfully with anticonvulsants, and—rarely identified, but
probably more common than we realize—hypersensitivity to
certain foods and allergy to preservatives and artificial gel-
ling agents in canned food undoubtedly exacerbates, if not

causes, some problems. We wait for science to catch up with our observations and explain more of the effects of diet on human and animal behavior, and in the meantime try to eliminate this possibility as a cause of aggressive behavior in cats and dogs by feeding consistent, balanced, and preferably fresh diets, though the higher-quality dry diets can be more helpful with canine aggression cases. Canned foods are out immediately, however.

Ostensibly incurable, motiveless aggression in cats may in some cases turn out, on postmortem examination, to have been caused by certain types of brain lesions, perhaps a tumor or from scar tissue resulting from an infection. This is especially suspected if the lesions are discovered along the midline of the brain and in certain areas of the hypothalamus, where electrical stimulation of live cats invokes an aggressive response. However, this does not help the live pet cat and its owners very much, at least not until brain surgery of cats becomes a viable approach to treatment.

Treatment of cats like Gong can, however, be very successful even if the cause of the aggression cannot be determined accurately. According to some animal ethologists or behaviorists, aggression is a kind of genetically predetermined life force, the periodic expression of which is almost obligatory from time to time to keep the animal rational at other times. Other theories suggest that it is only the capacity to be aggressive that is genetically preprogrammed and that certain stimuli are required for actual aggression to be released and that for the most part its direction will be toward that stimulus. In the light of either theory it is logical to suggest that a highly intelligent self-determining creature such as a cat requires a certain amount of stimulation, both mental and physical, for the release of energy and emotions. Never is this more true than with the Burmese breed, which is renowned for its intelligence and demanding nature. It is not surprising that it will take very little to excite the cat that is kept inside in a consistently unchallenging environment, however comfortable and well fed the cat may be. Against a consistent backdrop of soft, unstimulating comfort the cat will be all over the owners when they come home because their arrival is the major change of its day. All its energies will be directed at them at this time and any other time when they care to make a fuss over it. But Gong will also tend to overreact to other stimuli that most cats would take no notice of. Clare and Steven's entry into the room is such a massive change in the norm that Gong overreacts in a major manner with hardly any assessment of the situation and certainly no recognition of his owners until he has inflicted serious injury. Once the energy has been released a little, the cat will often perceive himself to be at risk with his teeth locked onto his owner's leg and will run away just as quickly. He may look confused afterward and engage in some of that unnecessary but tension-relieving intensive grooming.

Treatment is quite simple. We have to provide much more stimulation in the cat's life when the owners are not there so that this situation does not arise. Ideally the cat should be allowed to go outside, and once I have assured the owners that they are possibly being cruel by keeping the cat indoors, most who can, agree to do so and are never attacked again. The cat gets all the stimulation it needs through hunting, reacting with its environment and other cats, and returns home for food and affection.

This is not always possible for apartment dwellers like Gong, though I have known of owners who built very clever access ramps zigzagging down the outside wall for two floors to allow their cat access to the yard. Other owners do not wish to expose their cat to the dangers of the outdoors, and in areas with busy roads this is entirely reasonable. For these cats we must provide plenty of other forms of stimulation, and this comes most readily in the form of another cat. Properly introduced, the company and sheer presence of the other cat can relieve all those emotions in the aggressor and make life safe for everyone. Interestingly I employ a similar technique when treating feather-plucking parrots. This is a stress reaction to boredom. Providing another parrot in a cage alongside the sufferer is often all that is needed to alleviate the boredom and cause the feather plucking to cease. In most parrot cases the two birds hate each other, but simply having competition to scream at and keep a beady eye on is enough to keep the feather plucker busy. With aggressive cats we hope for, and usually get, at least a tolerant relationship between two cats, with occasional play and the odd disagreement all being part of the enriched scene.

For owners who can't or won't take on a second cat, I suggest that they get their cat accustomed to wearing a harness and being taken out for walks like a dog at least twice a day. The cat should also be offered as many novel objects as can be found every day. They need not be expensive; simple cardboard boxes, branches, toys, etc., all keep its curiosity exercised, and the owners are also advised to play intensively with the cat for an hour a day by making him chase moving toys in simulated hunting activity. But in many of the cases of solitary, confined, and rather unstimulated cats their general activity level can be managed from time to time with low doses of Valium or progestin treatment. Even better is the use of alternative treatments like Bach's Flower Remedies, a series of homeopathic medicines available in most health-food stores. But always at the back of my mind is the worry that certain cats need to be offered the outdoor life to preclude the emotional buildup that leads to outbursts of aggression, and while we can substitute, stimulate, and moderate, what is really needed is for the cat to be put in a home with a more fulfilling lifestyle and the owners to try again with two individuals of a less assertive breed perhaps better suited to life as apartment cats.

Redirected Aggression

Dear Mr. Neville,

Tyson is my big, bruising tabby cat. He is a neutered boy but nonetheless is always fighting. Thankfully he usually wins and chases every other cat away from the garden. But sometimes he will sit by the patio windows and simply watch over his territory. If another cat walks through the garden or even along the fence (why they still do it, I don't know), Tyson becomes extremely agitated. He growls like a dog or meows, his mouth trembles, his hair sticks up, and his tail swishes angrily from side to side. I can understand why, as he hates all other cats being on his patch, but when I try to comfort him he attacks me immediately. Last time he bit my leg so badly that I had to stay in the hospital for over a week. But I love him dearly despite this because he is usually very affectionate with me, when he recognizes who I am! What can I do about his aggression toward me?

Yours recovering,

Marina Sargeant

If it's not going to be a Burmese, it'll be a tabby! While there are no real data on the disposition of any color types toward certain behaviors, it does seem that while most tabbies are gentle and affectionate, despots like Tyson do occur, and if there is going to be a single dominant tyrant in a feral cat colony, in my experience at least, it is nearly always a tabby. Old English Tortoiseshell colors (that very attractive ginger and black marbling found only in female cats) are renowned by many cattery owners for being spitty and not easy to handle, while many claim that black or black and white cats tend to have friendlier natures than the average cat.

Tyson's behavior is a classic example of redirected aggression and, in this case, a very dangerous example. The excitement buildup at the sight of a rival daring to parade in front of Tyson has no release because he is frustrated in the desire to chase the other cat away. Tyson is fixated and studies every move with such intensity that he does not, as would be expected, immediately go to the back door to ask to be let out, as he would when needing to relieve himself.

Instead he sits and fumes. His owner, seeing his frustration, tries to comfort him, but unwittingly her approach triggers the release of his aggression, perhaps because he hasn't noticed her behind him and is momentarily startled by her, perhaps because her movements alone tip his aggression over that threshold of control. The severity of his attack is unusual, and while the immediate injuries from teeth and claws will be severe, the stay in the hospital was required because of the likelihood of cat bites and deep scratches becoming infected. Teeth and claws are coated with all sorts of unpleasant bacteria, but particularly the Pasteurella species, which when injected deep into our tissues can cause septicemia. A long course of antibiotics and rest is required to fight off what can be a very nasty reaction. This infection of bacteria is also the reason so many injuries to cats from fights with other cats turn into those purulent abscesses. But if it's any compensation, in the U.K. at least, there is only one mammalian bite loaded with more potential aftereffects— not a dog's, not a rat's, but that of a human.

As for Tyson, greater access to the outdoors and encouragement to spend more time out there by installing a cat flap will help ensure that his aggression is directed toward his rivals. Keeping the curtains drawn over the patio windows and other lookouts to block his line of vision may help, and offering a warm, attractive bed well away from windows may direct him away when indoors. Most important is that Marina not seek to calm him physically when agitated. Instead she should employ distraction techniques from a distance by rolling a ball of paper or foil past him to encourage chase behavior, by rattling the treat box, or by opening the back door and calling him, in order to develop that desire to go out and protect his territory at such times. Marina shouldn't walk too close, and if distraction is impossible she may be wiser to sneak quietly out of the room by the nearest unblocked exit. A low-dose prescription of tranquilizers from Tyson's vet may keep him less reactive, but in the long term alternative remedies may be preferable. It will also be worth testing the effects of diet on Tyson's reaction threshold by weaning him off premium-quality canned food for a week or two and offering fresh chicken or fish instead. However, this is really to extrapolate the occasionally observed effects of diet sensitivity in some aggressive dogs. It may make little difference to cats.

Territorial (Dominance) Aggression

Dear Mr. Neville,

My boyfriend, Martin, has two Siamese cats that live permanently indoors with him in his apartment. One cat, Suki, is placid and adorable; the other, Minkey, I have never really trusted. He has always eyed me rather carefully, as if waiting his chance to attack, and I have always felt uneasy on my own with him. He got his chance this weekend. I got up in the night and walked naked out of the bedroom to use the bathroom. He was sitting in the narrow passageway, and in the half-light I stopped to wonder whether I should walk past him. I decided to proceed, thinking that it was only dogs who guard places. I was wrong. I had taken two steps forward when Minkey sprang at me, leaping onto my bosom and biting and scratching me severely. I screamed in terror, and he ran away. Sometime later my boyfriend appeared but he refuses to believe that I did no more than walk toward the cat. I had to go to the casualty department of the local hospital and have been on antibiotics ever since. My left bosom is badly bruised and has two distinct cat mouth–shaped bite marks on it. Needless to say, I am refusing to go to his apartment again until Minkey is reformed, rehomed, or preferably destroyed, and setting a date for the wedding is unthinkable.

Yours sincerely,

Carol Summer

Such directed aggression toward people by cats is rare indeed; in fact this is the only case I have ever been called on to treat. Other aggressive cats of this nature are probably not given the opportunity to repeat their performance, and it is true that most men in Martin's position would be faced with the "it's me or the cat" ultimatum and choose the former without delay. However, Martin was a little less keen to go through with the mechanics of parting with Minkey, and he thought that in time the cat would accept Carol's arrival. He may have been right. However, following the attack Carol naturally became ailurophobic. (Ailurophobia is a fear of cats, *ailouros* being ancient Greek for *cat*.) So it was no longer just a case of trying to get Carol friendly with and accepted by Minkey but also a case of overcoming her fear or reluctance to be near any cats at all.

Ordinarily I would have advised that Minkey be put in a different home and that Carol try to make headway with Suki. But in this case it seemed that Martin was not prepared to countenance such a loss of bachelor ties with a faithful friend, and so Carol decided to break off the whole affair. Then we had a lucky break. Martin moved out of his London apartment to the Hetfordshire countryside, and the prospect of allowing Minkey free range outdoors became possible. In his new home Minkey became far less tense and far less attached to Martin, treating him more normally as a provider of food and as an occasional lap to sit on rather than as the center of his existence. Monopolization of key areas of the house, and in particular narrow stairways, became far less important to Minkey. In his old apartment Minkey had been behaving in a very doglike fashion. More than that, he had been behaving like a very dominant dog and at certain times expected the right of control over pack mates as to where and when they moved around. I treat many such dogs that guard the passageways of stairs, corridors, doorways, and, at night, the entrance to the bathroom. With no room to walk past and maintain a respectful distance, innocent people approaching and trying to pass must come within the animal's personal space. While dogs usually growl to force a retreat, Minkey simply leapt up at Carol as soon as she made it clear that she was continuing her advance toward him. The consequences were extremely painful and unpleasant and occurred as a direct result of the confrontation over the right-of-way in a narrow enclosed area. Unusual, even unique, but not worth the risk unless the cat's desire to dominate and maintain rank could be dissipated.

This could really be achieved only if his home could become a sleeping and feeding lair for Minkey, and this was the result of the move to the country. Alas it did little to overcome Carol's fears, but with new agreements regarding Minkey not being allowed in the bedroom, being shut in the kitchen on his own bed at night, not being left in a room with Carol alone, and using a protecting indoor pen for him to allow safe close contact with her, the couple got back together.

Social Conflict

Dear Mr. Neville,

Recently we bought a little kitten and brought her home to meet our two-year-old neutered, very affectionate male cat, Bertie. We tried to fuss over him a lot during their first meetings, but even after many weeks of trying, Bertie will not accept Suzy. We have to hold him back to prevent him from attacking her, and he has even scratched us in his efforts. On the couple of occasions that he has gotten to her he has bitten her neck and grabbed her with his paws while kicking her quite severely with his back feet. On one occasion, having grabbed her neck, he began to mount her, which would seem to be the opposite reaction. Is Bertie confused, and is there any hope of getting him to accept her?

Yours sincerely,

Jenny and Julian Wright

Socializing a new kitten or cat with existing cats in a house is one of the most common problems that I am contacted about. It is also one of those problems where it is very difficult to predict whether treatment will be completely successful because every cat is different regarding what others, if any, it will allow to share its home and to what extent. There is always the presumption that all cats can and should get on. After all, many people have nine, ten, eleven, and more, and they all get along quite happily and readily accept newcomers. Yet, like Gong of the misdirected aggression case, there are always some who simply will not tolerate another cat in their outside territory, let alone indoors. It all comes down to the key to the cat's success as a species: solitary as a hunter and unable to cooperate with others to coordinate a hunting strategy yet retaining the ability to share resources and avoid social conflict when everything is plentiful. So why, when we provide everything in plenty, from love and protection to bowlfuls of food, shouldn't a cat let others in?

As we might imagine, there are several important influences to consider. The cat that has been kept with lots of others since it was a less competitive kitten and has been used to meeting and sharing resources with others is more likely to accept a new arrival and tolerate it well after an initial introduction period. Full acceptance and friendship may take some time while the newcomer acquires the communal smell of the resident cats, dogs, and owners. The resident cat must also have time to weigh how the owners and other animals treat the new arrival; their acceptance and relaxed behavior may help its own. In many respects kittens are more likely to be accepted than adult cats because they are not sexually or particularly socially competitive. A bite on the tail from a kitten is like a kick in the shin from a three-year-old child; although unpleasant and worthy of rebuke, it is not nearly as challenging as exactly the same behavior from an adult. Having newly introduced young cats castrated or spayed as soon as advised by the vet will also help to reduce their competitive profile. I well remember the case of a marvelous family that I saw at my clinic at Rutland House Veterinary Hospital in St. Helens. They owned a beautiful young ginger tom that, after being accepted by the original black and white three-year-old neutered male cat, had suddenly and seriously been attacked by him. The advice was straightforward: have him neutered without delay. Sure enough, relations were restored between the cats in a matter of a couple of weeks. It was simply that Ginger had reached sexual maturity to the point where his male hormone level, or more importantly the pheromones (sex smells) that this caused him to produce, were too challenging for Blackie, who then felt that this intruder should be expelled.

Not all kittens are acceptable from day one, so it is important to introduce them carefully to the home and resident cat. Yet again an indoor pen is invaluable for this. The kitten can be housed inside with a warm bed, litter tray, food, and water and will be protected from any initial effort by the resident cat to drive it away. The resident cat(s) and dog(s) can then grow used to the presence of the new arrival and its smell within their home base. Ideally kitten and pen are moved from room to room over the course of a week to establish a right of presence throughout the house. Damp contents from its litter tray can also be placed in the original cats' trays to help in the smell-sharing process, and the cats can be fed alongside each other, separated by the bars, to encourage the sharing of personal space. The first free introductions should be supervised carefully, with one or both restrained by hand or on harnesses. Usually the kitten, having come from an environment of sociability with its mother and littermates, will be quite unafraid of the residents' approaches and sniffs and so will not fuel any adverse interest by running away. Indeed when I was using this technique to introduce my own Siamese kitten, Scribble, to the resident adult male cat, Bullet, and three large dogs, it very quickly became apparent that the cage was for their protection, not hers. From day one of freedom she occupied the 126-pound bullmastiff's bed and warns her off with a stare if ever the dog has the audacity to be in it when madam feels like a rest! The same techniques can be employed to introduce adult cats to each other, though the process may take longer. Some go as far as rubbing used litter from one cat onto the fur of the other, and it seems that this can help in some cases if the owners can bear it. Use rubber gloves, however!

Bertie's reactions to the new kitten are excessively aggressive; because of his immediate overreaction, there is little chance of his learning to interpret the kitten's arrival as nonthreatening. The attacks are obviously designed to force her to leave, yet the mounting behavior is also far from the amorous gesture we might imagine. Sexual postures are not always what they seem, and in this case they are a clear attempt by Bertie to dominate the kitten in the same way as more social animals such as cows, rabbits, and dogs use them to define dominance orders. Also, the areas of the brain that activate during sexual arousal are close to, and in some cases the same as, those involved with certain types of arousal during aggressive behavior. The overall reaction will clearly frighten the young kitten, and as is Bertie's aim she will be unwilling to stay around. Sex does not actually play a role in this behavior, though it is interesting to note that sexual responses such as mounting do readily occur in dogs and occasionally in cats when already aggressive or excited for other reasons. Mild sedatives, homeopathic or traditional, for the established resident have been known to ease the acceptance of new arrivals where pronounced and enduring aggressive reactions are encountered, and also for new adult arrivals particularly upset and defensively aggressive toward the inquiring sniffs of the resident.

However, there will always be cats that will not accept the presence of others at all and respond aggressively at every sighting. This prospect increases with age but decreases with the number of cats owned. There appears to be a dampening effect on social aggression when the would-be aggressor is surrounded by familiar cats. To respond aggressively to any newcomer is presumably to draw attention to yourself and risk that movement- or excitement-elicited attack yourself. Hence, even in overcrowded cat pens in animal shelters with apparently little room, newcomers may be accepted with no more than a sniff and a small hiss, but a big readjustment of each cat's position as they all move around to maintain as large a personal space (flight distance) around themselves as possible is likely.

With pet cats it is essential to be reasonable in trying to introduce new cats, young or old. About three weeks of controlled introductions, with drug assistance for adults if required, is sufficient time to indicate whether the dust will settle and harmony is likely to be achieved. But after this period it is essential to face up to the fact that the original occupant will not be particularly happy with having to share his home with this new arrival and that the new arrival may always live in fear of the original occupant and really never develop her own character to the full. In this case the rule of "last in, first out," should apply, irrespective of how cute the new kitten looks, and a nice new home should be sought rather than persisting too long, especially with kittens, for whom being in a relaxed, friendly home environment while still young is crucial. Having said this, I have several cases on file where, having tried my hardest to get cats to accept each other and reached the decision to place the latest arrival in a new home, the owner persisted for one more week or so and now the cats groom each other and sleep, eat, play, and go out together perfectly happily, which just shows how variable cats can be and how much we have yet to learn about their social habits.

Territoriality

For all of us with outdoor cats their territorial defense and lack of sociability with most other local cats is something we take for granted. While some do manage to make friends with the cat next door and some even establish very strong pair bonds with the moggy from the next block, the majority are intolerant of others in their patch when they want to use it. Neutering reduces much of the motivation from tomcats, especially, to be territorial and enables far higher densities of cats to coexist relatively peacefully in tightly packed suburban neighborhoods. Their ability to accept others once the need to maintain a territory that defines a level of food availability has been removed is a pointer to the elasticity of cats' sociability. By neutering them we also remove much of the hormonally motivated obligation to carve out a kittening area by females or a territory to encompass those of several females by males, and much of the requirement for social contact for reproductive purposes, such as queens in season calling for mates, also disappears. The pressure to manage any territory is largely off, yet some cats remain great defenders of their patch. In highly populated areas it seems that there may not be enough of a patch for everyone, and the dominant or most territorial individual will find that it needs to be on constant patrol to repel all intruders.

Cats like Tyson are typical examples, but most surrounding cats will gear their usage of his territory around his absence. When he is indoors and asleep, other cats will move through, aware of the reduced risk of expulsion from the staleness of Tyson's smell at his spraying posts and from having learned that he is regularly absent at certain times of day and night. Tyson will probably be most anxious to patrol at the key times of dawn and dusk, the usual hunting times when most prey is active, and so as a dominant cat will have established unopposed right of access to his territory at these times. Then other cats may flee without putting up any opposition so as to afford him his rights and to avoid conflict, yet at other less crucial times expect to be able to use the territory and walkways without hindrance. Tyson, or perhaps a slightly less assertive cat, may even be prepared to watch them walk through past his nose. The system has become one of time budgeting over a limited desirable resource, in the same way as we rent a week per year of an attractive vacation condominium, and is yet another remarkable adaptive feature of the social life of the cat.

After many years of being ignored by biologists, the behavior of the cat, both pet and feral, has become the feature of much scientific interest in recent years. Dr. David McDonald of the Department of Zoology at Oxford University, previously known for his marvelous studies of foxes, has been studying the social life of free-ranging farm and feral cats for a number of years. One of his students produced an enormous Ph.D. thesis on the social interactive behavior of feral cats that has done much to stimulate the imagination as to how and why cats communicate and what affects their social organization. Dr. John Bradshaw of the University of Southampton has also been looking at the social interactions of feral cats, and the Universities Federation for Animal Welfare at Potters Bar, from whom I had the honor to be employed as research biologist for three years, has also sponsored much research into the behavior and, where necessary, the control of feral cat colonies. The media too have started to become interested in the tiger in our living room, and there have been several very interesting television documentaries on the domestic cat in recent years. We tend not to see the cat's territorial activities as problems. What Thomas does outdoors is his affair, and as we can't control or train him anyway we shouldn't worry. Unless . . .

Despotism

Dear Mr. Neville,

Our two-year-old Burmese cat, Lurk, is a total hooligan, or at least has become one in recent months. For a neutered male he is now quite the most aggressive cat either of us has ever owned or seen outside of a wildlife documentary, though previously he was quite a normal cat. We have had endless complaints about him from all our neighbors, some of whom don't even own a cat! He not only chases and fights with all other cats outside in the neighborhood, but he also

seeks them in the safety of their own homes by
climbing through open windows, cat flaps, and
open doors! Once discovered, he attacks them
severely, leaving blood and fur strewn all over the
house and pausing on occasion to eat the resi-
dent cat's dinner. If the owners are present and
try to break up the fight, Lurk is quite likely to
attack them, and he has also been known to
attack young children playing in the yard. Our
neighbors are insisting, one of them via his law-
yer, that Lurk be destroyed, and frankly we agree
with their sentiments. Yet with us Lurk is ex-
tremely affectionate, so we are reluctant to have
him destroyed if something could be done to
build upon this home character. We have tried to
keep him in, but he paces and cries to go out and
is clearly very unhappy. Suggestions needed ur-
gently, please!

Yours sincerely,

Carrie and James Stroud-Parker

Such widespread despotism is fortunately far from common
in cats, though I should say that every case I have ever seen
has concerned a neutered male Burmese of between one
and two years of age. Perhaps this indicates some breed
disposition to despotism, though there may be many other
despots of other breeds and lineage that have been destroyed
for their Pol Pot approach to others rather than being referred to me. Such a constant expression of aggression toward all other cats with no respect for the security of their
home base, redirected aggression toward any protective
owner unwise enough to intervene, and targeted aggression
toward children would lead the majority of rational folk to
believe that the state of Lurk's mind is severely disturbed,
perhaps by some form of brain tumor. While this may be the
case, it is equally possible that Lurk is basically a tough,
dominant character who has learned the success of his own

strength. "Absolute power corrupts absolutely" as the saying goes, so having steadily achieved more territorial control over neighborhood felines, Lurk has decided to exact the tax of total right of access to their shelter base and food bowls.

Treatment can be surprisingly successful but involves enormous amounts of cooperation with the locals. Essentially Lurk has to meet with failure and rejection for a few weeks with regard to his relations with other cats. First Lurk is taken to the vet for a course of progestins, feminizing hormones with a calming effect, or a longer-lasting injection of essentially the same treatment in a stronger form. Then his patrol hours are rigorously controlled by his owners to coincide with times when all other local cats are safely shut in behind closed doors (and windows). At other times Lurk is placed in a pen outdoors where other cats can overcome their fear of him and approach with more confidence. Lurk's efforts to repel them or engage in warfare meet with failure, until gradually they achieve rights of occupancy in his territory and he comes to accept them more. Owners of the other cats enforce their pets' defense of home by hosing or throwing water at Lurk whenever he is seen in their yard, and their cats are fed in a totally secure region of their house, without any food being left outside that might encourage Lurk. Finally any possible influence of diet on Lurk's behavior is removed by providing a fresh chicken/fish diet for a month or so in short feeds to keep him coming home. Controlled excursions into the garden with Lurk on a harness and lead are also carried out, and any attempt on his part to be aggressive toward other cats is met with a jet of water in the face as a negative conditioning technique to build on what associative learning processes may be present. While this treatment, which is commonly employed in restructuring aggressive antisocial behavior in dogs, may appear contradictory to the earlier comments about cats not usually associating such unpleasant consequences with their actions, in some cases, when timed accurately, it can help in treating aggressive Burmese, Siamese, and other oriental breeds. Perhaps they are more doglike than we realize! The treatment above clearly demands much from owners and neighbors but has proved successful even where full cooperation has not been obtained. Such success has surprised me enormously, but perhaps we should view despots like Lurk more as school bullies than deranged cats. A good dose of embarrassing defeat seems most reformative!

Lurk is by no means unique in his despotism, but imagine, if you will, the disruption such a character could cause in a group of cats that live permanently indoors.

Dear Mr. Neville,

I have to confess to being obsessed by cats; indeed I have twelve of them. They are all quite peaceful except for Boris, a neutered male tabby. He tolerates, without being friends with, all of the other eleven except for Gemmy, a spayed female. To say he hates her is an understatement, as he spends his entire life indoors stalking and attacking her. She is already rather a nervous cat and naturally is very upset by Boris. She slinks around and hides behind the sofa for long periods, and though we try to keep them apart, doors get left open sometimes and Boris is in and chasing after Gemmy before we realize. The ensuing fight frightens all the others, who run like crazy in all directions, some through the doors, others out of the windows, others even going up and over us onto the dresser. It takes ages for life to return to normal after a successful attack. Is there anything we can do? Poor Gemmy has had so many wounds and trips to the vet and really doesn't deserve this treatment from Boris.

Yours faithfully,

Karen Black

I have to say that total peace among twelve cats is rather a lot to expect, yet a friend kept nine Burmese without any problem other than a similar despotic relationship between Hugo, a tough male, and one of the females.

Despotism is observed in social animals although it is not usually the normal hierarchy or method of social organization. Normal canine society tends to be based on a linear hierarchy with the alpha figure (top animal) or one alpha figure from each sex being dominant, followed in sequence by the others. None are equal, and usually the order shifts only with the assimilation of adolescent pups into the pack or the expulsion of older members. Such a pecking-order system is observed occasionally in cat societies but usually only in high-density groups when food is in short supply.

More usually the cats' system is based on tolerance and shifting status. Even a very tough, dominant tabby in a feral colony may be unwilling to oust a weaker individual from a food item and will simply sit by and wait its turn.

Occasionally despots occur in both canine and feline groups where one warlord reigns supreme and all others have equal bottom rank beneath him (and it is usually a "him"). Other animals, despite frequent attacks for little or no reason other than the violent enforcement of superiority, seem to allow such domination. Standing up to such characters may be inadvisable, but the underlings prefer to remain alongside these animal Hitlers, presumably for the protection they provide against unknown challenges. Perhaps feline and canine despots have the same mesmeric leadership qualities that attract and bind followers to them as had Hitler, Genghis Khan, and all other successful human despots.

There are benevolent dictators, and perhaps this system of government is attractive because it reduces the level of decision making for the leader's acolytes. Fine, if you can find a benevolent dictator, as Gemmy's companions seem to have done with Boris. He presumably receives constant acknowledgment of his dominant position from them; they probably give right-of-way and access to the food bowl and can be moved from sunny spots and comfortable chairs simply with a stare from Boris. Not so fine if you are Gemmy, the scapegoat for all ills and used by Boris to exhibit the terrible consequences of challenging the despot's authority to all the others. I am reading too much into what may well be a simple case of incompatibility between two cats, but such problems can cause such enormous disruption to what otherwise could be a harmonious relationship among cats. Quite apart from all the others understandably panicking at the onset of Boris's attacks on Gemmy, the general level of tension in the house will be constantly high, and risk of injury to innocent bystanders, feline and human, is high. Not only does someone have to separate the combatants, others may be scratched and torn by frightened cats trying to get out of the way.

While it may be kinder to put Gemmy in a new home for her own well-being, my experience is that such a move simply causes cats like Boris to select another victim to hound. Finding a new home for Boris isn't so easy either— would you want him with your cats? In any case, he's probably quite a nice old tyrant when he's not attacking Gemmy. Owners usually have a liking for the Borises of this world and may have a sneaking suspicion, largely justified in many cases, that relations among all the others are maintained only by the iron fist from above and that real infighting could break out on all fronts if he were to go.

As before, our aim in treatment is to allow controlled exposure between Gemmy and Boris so that Gemmy's fear reactions or efforts to escape are denied and thus don't exacerbate Boris's chasing instincts. He gets used to accepting her presence as he has done with the others, and his attempts to chase or battle her away meet with failure. I commonly recommend the controlled use of sedatives for both sides in such socialization programs, withdrawn steadily over four to six weeks as introductions develop. Progress is usually surprisingly good, though there are inevitably a few hiccups and temporary reversals, and though Gemmy and her ilk never become particularly confident cats they are at least safe from attack. In many cases they possibly wouldn't be any more confident elsewhere and may even be worse off and more vulnerable as solo cats, even if the owners would put them in a new home.

I have mentioned the adjunctive use of alternative medicine in the treatment of aggression quite frequently in this chapter, and never have I encountered more success with this approach than with treating despots. Four drops of the Bach's Flower Remedy, Vine (available in most health-food stores), diluted with water and squirted into the cat's mouth via a plastic pipette two or three times a day, if it won't be accepted in drinking water or milk, has produced some amazing results. Despots like Boris under this treatment are still arrogant strutters, but instead of attacking their usual victim, they simply ignore the cat or, at worst, hiss and walk away. I wish someone could tell me why this treatment is so effective in these cases.

Throughout this chapter we have looked at the dangerous side of our pet cats in relation to the cats' normal prey, other species, other cats, and ourselves. With the exception of predatory aggression and territorial aggression toward other cats, thankfully all are rarely observed among the feline population and are to be found in concentration only in my case files. Two important points emerge. One is that aggressive cats should never be tangled with, as once excited to the point of fearful or assertive aggression the usually discretionary cat is well armed to inflict serious injury on us. Secondly, when already aggressive it will also be activated enough not to recognize us as friends or interpret our intervention on its behalf as anything other than threatening. Where intervention is essential to interrupt aggressive behavior, it must be via buckets of water or the use of startling devices such as rape whistles, never with hands or feet or by advancing into the fray. Such intervention may decide the battle but set back the winning of the war, which must be tackled more logically long-term looking at all the influences on the cat's behavior. Thankfully most problems are resolvable given sufficient time.

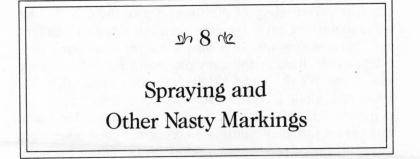

৯ 8 ৫

Spraying and Other Nasty Markings

The majority of the cases I see concern cats that mark their territory indoors. In fact you could say that cases of indoor spraying, furniture scratching, and, worst of all, middening, are my "bread and butter." Middening is the deliberate act of defecating in the open to emphasize occupancy of part of the cat's territory. This is the most extreme and unpleasant type of marking but is in the same context as all kinds of other marking behaviors that our cats perform, most of which are acceptable or even encouraged. When a cat rubs around our legs or offers its body to be stroked, it is actually transferring some of its own personal scent onto us. The scent is produced in glands all over the cat's body and especially along the tail (caudal glands) in mature cats, from where it becomes dispersed more evenly over the fur when the cat grooms itself. The scent is then ready to be rubbed onto furniture, another cat, or us. This helps the cat develop a communal smell with us and its home so it can identify the strength of its occupancy at the core of its territory and also recognize us as something of a protective mother figure. Scent-producing glands are particularly concentrated under the chin, on each side of the forehead (temporal glands), and on the lips (perioral glands), which helps account for the enthusiastic head butting and facial rubbing that cats enjoy with us. Saliva is also produced to a greater level during such interactions to help spread the message.

The presentation of the underside of the chin by the excited, purring cat is far more than just a display of affection between friends. It is also a way of cementing that relationship through the cat's anointing us with his own scent, and it's the way friendly cats greet each other. By presenting their faces and rubbing each other they each acquire a little of each other's scent to cement their bonds and ensure that they continue to recognize each other after any short period of separation. The activity of these glands is reduced when male cats are castrated, and this accounts for much of the loss of that tomcat smell. Perhaps a cat's personal scent is also altered when it is traumatized, for it could be that the recognition system suddenly breaks down so that an accepted friend is no longer identifiable and is repulsed in the way described in Chapter 6. We don't know, nor do we understand very much at all about the importance and method of scent communication in cats, or any other social animals for that matter. It simply isn't our medium, and though we have the sense, we don't have the senses to detect what is going on.

We can, however, see where the cat combines scent marking with a visual signal at scratching posts. Most of these posts are outdoors in the competitive zones of yards and countryside, but occasionally cats will need to scratch indoors too. More than just keeping claws ready for action by sharpening on chairs, curtains, and favorite of all, burlap wallpaper, the cat is also leaving a definite scent mark derived from glands situated on the sides of the pads. These glands also help the cat mark out walkways indoors and out, both to identify the cat's usage of the path to other cats and to help the cat navigate its way home or to favorite activity areas or sleeping and feeding lairs.

Scratching

Dear Mr. Neville,

Sidney is a lovely cat, but he will insist on scratching my best furniture at the slightest opportunity. My sofa is shredded, and many of my chairs are in rags. I've tried telling him off, but it only discourages him at the time. He simply returns later to have another go. Is there anything I can do?

Yours sincerely,

Samantha Swift

The first thing we have to do with this type of problem is decide just how much of the scratching is likely to be of a scent-marking function. Usually the cats that scratch a great number of areas in the house, especially around the access points near doors, are marking their territory with scent and the visual sign of the scratched surface. These must be treated in the same way as indoor sprayers, and the most important feature of treatment is finding out why the cat feels the need to mark what should already be secure. Cleaning up the marking posts and controlling the cat's access to them for a while, as well as improving its perception of home as a safe base, are also important, and we will take a closer look at this later in the chapter.

When the problem is centered on one or two scratching posts only, the emphasis is more on keeping the claws in shape. Scratching doesn't actually sharpen the points on the claws; in fact it causes the whole outer layer of the dead claw to slough off and exposes a new, pointed claw tip underneath. It is often easy to find old claw outer layers discarded on scratching posts in this way. Cats that scratch indoors may be doing so to avoid drawing the attention of local rivals outdoors, but usually it arises through a lack of transfer, as a young cat, from using available surfaces at home to the more acceptable outdoor surfaces of trees and fenceposts. Treatment is therefore aimed at trying to redirect the cat's claw-sharpening activities rather than punishing it for doing what is needed naturally to keep the weapons fit for hunting. Placing a suitable alternative of a type of material that is not found elsewhere in the house on or directly in front of scratched areas usually does the trick. A large piece of bark or one of the commercially available sisal or string-wrapped posts or cards can easily be positioned at the scratching areas and, once the cat has begun using it, can be moved a few inches at a time to a more convenient area nearby in the room. Using a bark alternative may even help the cat transfer all scratching to the outdoors.

The need to scratch is obviously inherent in all cats as part of keeping in shape, and so a suitable scratching post should always be made available to young kittens that are to be kept indoors permanently. This is far more sensible than the outrageous approach to solving the problem by surgically removing the cat's claws. Thankfully, few veterinary surgeons in the United Kingdom are prepared to carry out this unnecessary mutilation except on medical grounds and especially not if the cat is ever to be allowed outside, where it would need its claws for climbing, hunting, and self-defense. While the adult declawed cats I have seen seem to bear no ill effects long after the surgery, I must say that the most miserable, healthy cat I ever saw was one in an American animal welfare hospital that had been declawed the day before. I only wish veterinary surgeons worldwide would take a strong stand against all mutilation of pets, including the docking of dog's tails.

Spraying

Most people do resolve problems of scratching indoors by providing posts or by teaching the cat that only one end of one old chair is acceptable. The majority of the cases of unwanted indoor marking that are referred to me concern the spraying cat. Spraying is a very deliberate act of marking by males and females and not to be confused with urinating, which is more usually performed from a squatting posture into a prepared hole in a movable substance such as soil or cat litter. Spraying is performed from a standing position, with the cat facing away from the feature to be marked. Typically the cat stands an inch or so away from a vertical object such as a fencepost or shrub and, with tail held high and quivering at the tip, directs up to ½ teaspoon of urine backward in short spurts, designed to spread the spray over the target area. The process looks quite difficult to perform, and perhaps the necessary muscular spasms around the penis in males account for the stepping action of the hind legs, the slight arching of the back, and the heavily concentrated look on the cat's face.

Scent marking is performed by many animals and is often readily detectable by even our poor sense of smell. Polecats, mink and ferrets, dogs, and mice all deposit very distinctive odors in and around their territory, but there is little to compete with the unforgettable smell of tomcat spray. Such smells are known as *pheromones*, which are secreted in a fatty viscous form from glands placed all over the body. Pheromones can simply be released into the air, particularly by social animals, but are specifically directed to concentrate scent on marking posts by many mammals. Glands producing pheromones are found on the abdomen, feet, and face of mice, and many species of deer mark grasses or shrubs with secretions from glands below the eye. Many mammals have their most active scent glands located in the genital region. Urine and feces are good vectors for distributing pheromones and so are anointed with smelly secretions from the accessory glands either during normal elimination (scientific hype for toileting) or by specifically targeted application as in spraying cats or leg-cocking dogs.

While insects such as bees and ants also communicate through pheromones, birds do not, and though many smell, they do not seem to use this as a method of influencing each other's behavior. With cats, scent posts are used to allow a longer-lasting signal to be given rather than simply releasing smell into the air. The signal may last hours or even weeks after its deposit and be detectable, given the right atmospheric conditions, up to fifty feet away. The importance of smell communication is easily overlooked, as the cat outwardly appears to be a creature evolved to rely on senses of sight and hearing; certainly these senses are employed most in a cat's relationship with us.

Males of scent-marking species usually possess more scent glands than females and mark more frequently. Cats are no exception, and unneutered males spray more often and certainly more pungently than other cats. Few people can live with an unneutered tom (a lovely expression for the uncastrated male) who sprays indoors, which is the reason males are usually castrated before the behavior develops at adolescence, unless the lucky chap is to be a stud. In this case he will probably have to accept life in his own personal enclosure at the edge of the yard, with the compensation of occasional trips to the cat show to be admired, win prizes, and be signed up to perform with a series of attractive young female cats. Even if you smell, life can have its compensations when you're a cat.

Unneutered toms begin to spray at adolescence with the onset of male hormone production, as the result of the need, it is thought, to become socially active in procuring a territory and attracting females. Spraying frequency, together with the rise in testosterone levels, increases with the presence of females in season and during the feral or stud feline breeding season. Consequently the increased movement of toms means there is a growth in the number of competitive encounters with other males. Interestingly castration after the cat has started to spray causes an immediate cessation of the behavior in nearly 80 percent of males of all ages and slower diminution of the behavior over a few months in another 10 percent. Roaming and intermale fighting also decrease in the majority of cases. A few males continue to spray despite being emasculated, and a few that are castrated before puberty also start to spray sometime later. This indicates that while spraying is largely a secondary sexual behavior related to the presence of androgens, specifically testosterone, in mature toms other factors can also cause them to spray.

The smell and positioning of the spray at nose height to other cats helps the sprayer announce its presence to existing users of the yard or other territory. The spray is believed to act as a personal calling card, giving the finder all sorts of details about the sprayer. It probably indicates sex, rank and sexual status, state of health, and age, and the freshness of the mark will indicate how long ago the sprayer was at that point. A fresh mark may mean "beware," while a stale one may mean "relax and proceed in safety." Yet many cats simply investigate a sprayed area and walk away apparently unconcerned, without any modification of their behavior. They may overspray the mark, or they may not, and there seems to be little pattern in their response to the spray of an observed highly territorial individual from one day to the next. Perhaps spraying is like a present, more important in the giving than the receiving. Tigers, especially, spray extremely frequently if placed in an unfamiliar area, perhaps in an effort to surround themselves with their own comforting familiar smell rather than to announce their arrival to others. Feral toms too may enter a strange area where food is provided regularly and, when the area is quiet and no actual food is available, spray nearby to establish their smell as an envoy to resident cats. This is believed to desensitize the residents a little to the tom's actual physical arrival sometime later, when the food appears. His arrival is thus staggered, and as his physical entry is less of a sudden, alarming prospect to the resident cats, he is perhaps less likely to be ejected *sine die*.

However, it is hard to imagine that spraying is not intended to have some deterrent properties for rivals trying to compete for food, shelter space, and especially females. Hyenas have been observed to chase prey (they do hunt as well as eat carrion) as far as the boundaries of their territory. Once the quarry crosses into that of the neighboring pack, the hyenas stop abruptly rather than violate the border agreements. Instead they mark the border with urine, feces, and the associated pheromones and withdraw into their own area again. Frequent inspection of the border marks would reveal whether the rival pack had marked recently and was still in residence or had moved on, in which case they may feel safe to move in if desired. Generally mammals do not just mark the borders to their territory, but it appears that way with cats because fences are places where progress is slowed and encounters with other cats are more likely. Instead, cats and hyenas mark throughout their territory, perhaps endorsing the idea of the self-familiarization function of spraying observed in tigers. The more marks, especially fresh ones, that are encountered in a given area, the more the sprayer feels confident within it, but others moving through need not necessarily feel ill at ease. With the high density of pet cats in suburban and urban areas spraying posts are constantly marked and overmarked by a sequence of cats, depending on who is moving through the area and at what time. Doubtless the sprayed posts help define the time-budgeting system of territory occupation described in Chapter 7.

For any form of marking to be effective the message has to be understood by the receiver. Observing cats' responses to their own marks and especially those of other cats is enormously interesting because it involves a sense we do not possess. The flehmen (open-mouthed) grimace of the cat is also observed in other animals such as horses and cattle that possess the vomero-nasal, or Jacobson's, organ, a small scent-sense gland on the roof of the mouth. We, like all other primates, do not possess this organ and so lack this sense. When sniffing spray posts or urine the cat may seem to be in something of a trance, and with ears slightly flattened it wrinkles its nose and lifts its lips while drawing in the scent. The cat's tongue may flutter a little, and breathing may stop for a short while as the scent is drawn in through the gaping mouth and concentrated in the vomero-nasal organ, which is situated at the base of the nasal cavity and open to the mouth via two ducts in the hard upper palate behind the front teeth. The sense is therefore best described as midway between taste and smell and enables the cat to "taste" the full message of the smell under investigation. Sex smells elicit the flehmen response best, especially in unneutered toms, as does the smell of catnip with those cats that are sensitive or responsive to it.

So cats smell and smell-taste the messages left by others and to check on the freshness and content of their own marks. Just how much information they can gather this way is a mystery to us but is clearly of vital importance to the social life of the cat. Some behaviorists have even suggested that destroying the cat's sense of smell surgically by means of an "olfactory tractotomy" or, less drastically, by blocking the ducts leading to the vomero-nasal organ, will decrease the frequency of unwanted spraying in cats. Leaving the cat unable to detect the decay of its own smell, and with smell playing a far less important role in the indoor sprayer's overall sensory input, this may be a logical, if totally unacceptable, approach to tackling such problems. I would never recommend such surgery and would always rely on identifying the cause of the problem and/or managing the cat's access to that influence so that it learns to cope without spraying rather than interfering with a major natural sense. Sometimes I wonder what right we have to go chopping animals up in an effort to modify behavior problems. Castration, yes, for both our benefit and the cat's in most cases, but removing the sense of smell? Surely not.

Contrary to popular belief, it is not only unneutered males that spray. Unneutered females may spray when coming into season to announce their receptiveness to nearby males. They appear less likely to spray at other times, and one would expect them not to spray close to the nest when nursing kittens, to avoid attracting attention. But neutered cats, male and female, also spray their outdoor territories as a normal pattern of behavior. Many owners seem surprised at this and feel that perhaps their female cat is metamorphosing into a male or that their cat, male or female, is having difficulties in urinating and is being forced to adopt a standing posture for the purpose. The vet usually assures them that all is well, though if a cat looks in pain or discomfort when spraying outdoors or starts to spray indoors, it is always wise to ask the vet to examine a cat physically and carry out a urinalysis. Cats of both sexes can adopt a standing posture to urinate when suffering from cystitis, an inflammation of the bladder caused by injury or infection, or when suffering from a painful condition known as *feline urological syndrome*, or FUS for short. In this syndrome uroliths, or "stones," may become deposited in the kidneys or ureters, causing partial or even total blockage of the passage of urine. The syndrome was a frequently encountered problem when dry feline diets were originally introduced some years ago. The diets contained too high a level of magnesium as well as causing an alkaline urine, creating the ideal environment for precipitation of mineral crystals in the urinary system. Thankfully the formulas of many of these dry diets were quickly altered and one or two companies investigated the problem extensively before launching safer diets. However, as such problems may not arise just as a result of unbalanced diets, any sudden onset of spraying behavior in a previously nonspraying cat or difficulties at the litter tray should prompt an immediate visit to the vet. Caught early, FUS can usually be treated and involves close dietary management as much as anything else to prevent recurrence.

Virtually all cats, including neutered males and females, spray outdoors as a normal part of their behavioral repertoire. The higher the density of cats, the more they spray, and this is especially true of spayed females outdoors. As well as fighting more, assertive unneutered toms also spray more than other cats because territorial occupancy is most important to them, and so the need to impress a chemical presence on the opposition is perhaps linked to aggression. But neutered cats also spray more when crowded, without necessarily becoming more aggressive. Therefore, instead of viewing spraying as a dominant supportive gesture in maintaining rights to certain resources, we should perhaps view it as an emotional response to certain conditions or changes that includes, at the extreme end, arousal when aggressive or territorial.

Cats of both sexes vary enormously in their propensity for spraying, and the breeds that we recognize as more sensitive or intelligent, such as many of the orientals, seem especially practiced at the art. Cats may spray more when they are anxious or when vulnerable, perhaps because of a decreased level of attention from their owners. They may spray when their routines with us are disrupted or if they are punished or forced to share our affections with a new cat. For others spraying may be triggered by a combination of influences raising the cat's emotional state beyond a key threshold that is relieved partially by the act of spraying to surround the cat with a familiar security veil of smell. This includes those times when a cat may be aroused through the anxiety of defeat or else being victoriously assertive following an encounter with a rival. The spraying reaction threshold is apparently higher in neutered than unneutered cats and highest of all in spayed females. In short the nonreproducing female is perhaps the most competent, least emotionally reactive cat compared with macho toms, sensitive neutered males, or unspayed females, whose behavior is subject to a varying hormonal status. Certainly most of the cases of unwanted spraying behavior referred to me concern neutered males. It is also true, from my experiences and those of behaviorists in the United States, that while spayed females are less likely to be presented for treatment of a spraying problem, when they are presented they are more difficult to reform than males, neutered or unneutered.

Thus spraying is a natural behavior in all feline species and, for whatever reason or combination of reasons, has persisted because it bestows some evolutionary advantage on sprayers compared with nonsprayers. Or is it that more reactive, emotional cats, which are more likely to spray, are also more likely to sense and avoid danger and so survive better? Whichever the reason, spraying is perceived as a behavior problem in pet cats only if it occurs in our homes, and we disregard as largely untreatable the anointing of our shrubs or back doorways by free-ranging feral cats or unidentified local pets. Fortunately the majority of our pet cats do not spray indoors, because the core to their territory is secure from challenge by other cats and our pets are most likely to be emotionally calm there. We are there to protect them and behave like their mother, and they can rest without fear of disturbance. In the territorial sense it would also be a waste of time and energy to mark an area already defined as occupied and safe. In cases where many cats share a household, even apparently reactive or victimized individuals may not spray to avoid drawing attention to themselves and so enable them to remain in the home without being perceived as unnecessarily weak or challenging by other more dominant housemates.

So when a cat does start spraying indoors, it is usually an indication that the cat has been upset or challenged by some disturbance or change in our home that has made it feel less secure. Identify and remove the challenge and the cat's emotional state may wind down below the spraying threshold. The difficulty is usually the identification part, as we may be dealing with a cumulative effect from several influences, each one of which may seem relatively minor. The other difficulty is that the emotional threshold of every individual cat is different, and while we can identify certain gross triggers that would set any cat off on a spraying career, others seem enormously sensitive and start spraying in response to the most minor of changes. These are the most difficult to treat, as aspects of treatment itself may be sufficiently worrying to cause an initial upsurge in the frequency of spraying. Then there is the problem of not being able to catch the cat in the act of spraying. When we are with the cat, we act as a protecting influence, an emotional prop that allows the cat to settle and feel confident to the point of not needing to spray. So it may be difficult to identify when and under what circumstances the cat will spray its marking posts or to organize treatment or manage access to those points on anything other than a total exclusion basis.

If there are lots of cats in the house, it may also be impossible to identify the culprit or culprits if they are never caught in the act, in which case we may need to inject each cat subcutaneously in turn with a small amount of a liquid called *fluorescein*. This causes the cat's urine to fluoresce under the ultraviolet light of a Wood's light, which is more often used to detect ringworm. Fluorescence of fresh spray will indicate that the cat injected on that day is the target for treatment and all others *may* be innocent until tested. And if that isn't enough for a poor old cat shrink, there are also those cats that cope without spraying after enormous disruptions such as being beaten to pieces by the local feral unneutered tom on one day and then anoint every piece of furniture after being dive-bombed by a bird the next. Delicate souls, cats! Treat each one with individual care and attention!

Dear Mr. Neville,

We have enjoyed six trouble-free years with Bandit, our neutered male cat. He has followed us through three moves of house and always established himself well with local cats. He is not a great fighter but usually stands up for himself okay. Three weeks ago we came downstairs one morning to find that overnight our kitchen had been a feline battlefield. Another cat must have come through the cat flap to engage Bandit. We must have slept right through an enormous dust-up as there was fur everywhere. Most of it was Bandit-colored, so we presume that he got the worst of it. He bore a few scars but seemed unaffected in his general behavior, except for the fact that he now sprays all over the house. Not a piece of furniture has been left unsprayed, and he regularly makes the rounds and tops up his marks. This happens only when we are away, so he

obviously knows that it is wrong. Should we pun-
ish him? Our house is starting to smell most
unpleasant despite hours of cleaning up after
him. Please help!

Yours sincerely,
Sue and Barry Scott

If there is one single most popular invention for cat owners,
then it must be the cat flap, believed to have been developed
by the great physicist Isaac Newton, who, between having
apples dropped on his head, designed a small cat-size hole in
his door so that his cat could come and go as it pleased
without interrupting him. Had he but paused a little, he
would have realized that holes and, later, simple cat flaps
with hinged doors don't allow just the resident cat to go in
and out. All his friends and enemies can come in and out
too! Some pets are surprisingly tolerant of other cats coming
into the very core of their territory, stealing their food, and
resting in their favorite chair. Others never suffer such inva-
sions, perhaps because local rivals have never experienced a
cat flap and have no idea what joys lie beyond them. Then
there are those like Bandit, for whom home is always a castle,
wherever it is. Here he is protected by the family, loved and
cared for, and always has those safe four walls to retire
behind, away from holding his own against the local rivals.
However, the introduction of the simple flap has meant a
drawbridge is always down and any other cat can also come
in. If it happens to be the local unneutered tough feral tom
or the most hated of rivals from two houses away, we have
unwittingly created the most inappropriate battle zone.
Food, shelter, and affection, even from sleeping and nonpro-
tecting owners, are resources worthy of spirited defense.

Although ultimately Bandit managed to repel the intruder, this incursion into the safe core of his territory has
meant that Bandit now perceives the home as being under
the same challenge as the outdoor section of his patch. As
other cats can now enter at will, Bandit must try to endorse

his presence and occupancy chemically, by spraying all areas likely to be encountered by any other cat. Spraying will also surround Bandit with his own familiar smell, which is perhaps now more of a comfort than the communal smell of home, owners, and perhaps the other pets that he shares with. Since the invasion this shared smell is no longer sufficient to make him feel at home, and he must now rely on his own efforts. Emotionally this is about the worst thing that can happen to a normal cat. Small wonder that he sprays everything and tries to ensure that all marks are kept fresh. At first, as the owners clean up, the cat goes around reanointing the marking areas. Long after the original fight, and even if it occurred only once, a cat can be disturbed enough to keep on spraying indoors for months afterward.

If this sounds like a logical reason for a cat to start spraying, there are plenty of examples in my files of cats that have begun spraying indoors simply after the installation of a cat flap. The owners felt that they were doing the cat and themselves a favor, but all they did was destroy the important barrier between the outdoor competitive jungle and the safe indoor reserved zone. In some cases the cat immediately recognizes that the whole of his territory has become continuous and that there is no safe castle anymore. So if spraying is practiced and helps outdoors, then it ought to help indoors too. The cat is of course unable to distinguish between apple trees in the yard and antique Chippendale chairs in the living room. After all, they are both vertical and will hold the smell of his spray at exactly the right height.

It doesn't take a huge amount of intelligence to deduce that cat flaps are bad news for cats like Bandit and that if the indoor spraying is to cease the flap will have to be boarded up again so that security of the inner sanctum may be reestablished. Even so, the cat may continue to spray for a while and will need emotional mollification through extra love and affection from the family and perhaps a little short-term sedative or progestin treatment from the vet, as suggested for treating some cases of feline nervousness.

Cleaning

It is important to clean up the indoor marking posts effectively so that Bandit and his ilk do not go back to them to top up afresh over the decaying or stale smell of previous efforts. This alone can be a sufficient stimulus to prompt spraying, even though the original cause of the onset of the problem may be long gone. Unfortunately cats and people have different ideas as to what smells clean. To us the sterile wafts of chlorine or light ammonia denote that the area is as clean as can be, and we buy our household bleaches and many detergents on this premise. Sadly ammonia compounds are constituents of cat urine. To us the freshly scrubbed spraying posts of the table leg smells as sterile as a hospital room. To the cat it probably smells as if another cat has overmarked his mark. In the face of such a challenge the cat will simply over-spray that mark and so on ad infinitum!

Therefore, when cleaning sprayed areas it is vital to consider the cat's nose if we are to avoid further problems. Ideally the smell of the spray must be removed entirely without being replaced by a compound that will itself leave a residue containing ammonia, however clean it appears to our clumsy olfactory senses. Ideally too the smell of the cat's spray should not be replaced with the artificial smell of lemons, pine, roses, or any other of the standard scents added to cleaners and air fresheners. These scents will attract the cat back to its spraying post and inform it that something has been there and found the cat's mark sufficiently interesting to warrant leaving its own calling card. It may not be another cat, but it is something unencountered previously and clearly is presenting a challenge. Time to overspray again! Cleaning sprayed areas effectively to the cat's nose is not an easy task but seems to be best achieved by using a warm 10 percent solution of an enzymatic or biological liquid (or powder) laundry detergent to wash the area initially. This removes proteinaceous compounds in the spray. Then the area should be rinsed in cold water and wiped dry and then sprayed with a low-grade alcohol such as surgical spirit via a plant mister. The spirit should be lightly rubbed into fabric or wooden areas with a nail brush so that it penetrates the fibers or minute crevices where the spray may have reached. The spirit will dislodge any fatty residues that the detergent couldn't dissolve. The smell of the decay of these fatty residues in particular may prompt the cat to top up its own marks, so it is essential to try to remove them. A word of warning, however: this cleaning system may also remove dye from carpets, curtains, and other fabrics, and polish or varnish from furniture, so check before wading in by testing on an unimportant or hidden corner of the surface first. After cleaning, the area must be allowed to dry thoroughly, if necessary using a hair dryer on very wet areas, before the cat is allowed anywhere near its old spray post.

This cleaning system is usually effective against even the particularly pungent fatty spray of unneutered toms, depending on the accessibility of the cleaners to the sprayed areas. Alas some spray will often run down the vertical area it was intended for or seep deep into soft chairs, etc., and there is no other option for removing the smell but to rip up carpets or throw away chairs. My own clients' experiences indicate even the fruitlessness of putting chairs sprayed by unneutered toms into storage in the garage for the smell to wear off. It doesn't. But spray from neutered males and females is usually removable using the method above.

Deterrents

Removing the smell of spray indoors will at least help prevent topping-up spraying, but if the cat is still distressed or the cause of the distress has not altered, it will still be motivated to spray its marking posts. The same spray posts will continue to attract the cat because of their important geographical position in the house. Access points such as doorways and boundaries to the house such as windows are recognized as weak spots in the castle and deserving of extra identification. Any rival that gains access is most likely to encounter the resident's smell at these points and also on table legs and furniture close to the walkways a new cat might follow in the house.

So the aim is to try to deter the cat from spraying those frontier posts of curtains, window frames, doorways, and crucially placed table legs. There are countless products on the market said to deter cats from soiling indoors or in the garden. Like the more traditionally used ground pepper, chili powder, and mustard, they are usually unsuccessful. We are, after all, dealing with an emotionally disturbed cat, and to receive a nose full of pepper, chili, or evil-smelling chemical preparation at one's spraying post is probably only going to upset the cat more and cause an increase in spraying. While some claim success from tying aluminum foil strips to spraying posts on the premise that the cat is startled by the noise of the spray spattering on the foil, I have never found this to be effective. I have had some success by placing unpleasant walking surfaces, such as trays of marbles or pinecones, at the base of spraying posts as while standing on a movable surface the cat cannot adopt the spraying posture. However, in most cases the cat simply stands a bit farther away and sprays harder or moves along a little to another point. In other words, such tactics do not block the motivation to spray; they simply redirect it. Others have tried using remote punishment techniques of finely set mousetraps that go off as the cat goes toward the area and vibrates the trap with its footsteps; some have used an electrified foil wrapping around the spray post that when sprayed causes a small, startling current to run back up the spray and connect on the cat's rear end! Not nice and not even effective; as with ground pepper and all those smell or taste deterrents, the most that can be achieved is to transfer the spraying behavior elsewhere as they do nothing to alleviate or recondition the actual motivation to spray. As with aggression problems, punishment has precious little effect on feline behavior other than to cause flight.

Food on the other hand is usually an excellent deterrent to spraying, though it should be said that some individuals do simply move on to nonbaited areas. I have yet to encounter a cat that will spray its own food, presumably because it would make it distasteful to consume later. I say "presumably" because tigers are known to spray uneaten prey, having eaten their fill at a first sitting. They scratch vegetation and soil over the carcass and then spray it, perhaps to deter others from showing interest. This behavior has probably evolved in tigers because, although solo predators, they are capable of bringing down prey larger than themselves and certainly more than can be consumed at a single sitting. There may be enough left on the antelope joint to provide a cold meal the next day, so the covering and identification are perhaps acts of storing that are unnecessary with smaller felids, which have evolved to catch only single-meal-size prey.

This is just as well, because food placed on a saucer or in a little plastic container and placed at the spraying post will usually deter the indoor-spraying cat. Dry food is recommended for the purpose as it lasts longer, though it may need to be stuck down on the saucer to prevent the cat from eating it. Some cats I have known have emptied the saucer and then sprayed the post! Also the family dog, for whom stealing the cat's food is one of life's main purposes, may render the method quickly inoperative unless the food is stuck to a heavy bowl that is in turn stuck to the floor. Dry cat food seems to function as a deterrent to spraying even if the cat usually receives a canned diet and rejects dry food as part of its normal ration. The reasoning behind this is obviously not that the cat finds food unpleasant, as it would a nose full of ground pepper dust or an electric discharge to its nether regions, but that food itself is a reassuring influence in the core of the cat's territory. The core corresponds to its feeding lair, and perhaps wider dispersal of food in the area for a short period helps reestablish that core as being a secure feeding home base. Similarly flooding the sprayed rooms with objects familiar to the cat, such as toys, litter trays, beds, etc., helps to reassure the cat and negate the need to spray to achieve the same result.

Dear Mr. Neville,

After eight years of sitting in the same chairs we had when we were married, my husband and I

decided to splurge on a new suite. We also redec-
orated our living room to match the new suite,
but ever since we sat back to enjoy the new am-
bience of No. 28 Redford Road, the cat has
sprayed our new suite daily. We have been forced
to rescue the plastic covers from the garbage can
to cover the suite and protect it from Mick, who
has also registered his disapproval of our choice
of wallpaper by spraying that too in a couple of
spaces by the door. We may have to disagree with
Mick on taste and color, but how do we stop him
from spraying his opinions?

Yours sincerely,

Moira and Arthur Alexander

This type of spraying problem can hit hard at any family. To
the family the chairs have improved life for everyone, including the cat. He now has a bright, airy room and new chairs to
spread himself over if he's lucky. To the cat the very heart of
his territory has been totally altered. Furniture has been
moved, and all his walkways and escape routes are blocked
or simply not there anymore if a new carpet has been laid as
well. At cat's eye level the whole scene is unfamiliar, and it
smells strange too. Somehow the territory has been altered
completely and filled with novel objects, so to reassure
themselves that this is indeed home and to ensure that no
other rival moves in to try to occupy this virgin resource,
many cats will spray new furniture and wallpaper for a couple of weeks. Some are so sensitive to change within their
lair that they will be disturbed simply by moving existing
furniture around or adding a single new fixture. Indeed
many will routinely spray any new item brought into the
home, from black plastic garbage bags (always a favorite
target) to shopping baskets, cooking pots, and the children's
new toys. Some spray each new thing once only, but others
persist for months in their daily routine of identifying all
recently acquired chattels. It can take ages to make the

rounds, not to mention an enormous volume of spray. Other favorite targets are those where the cat's spray becomes altered after deposition, perhaps by warming. The casual squirt on the washing machine may even go unnoticed to us. Unnoticed, that is, until the washing machine is turned on, warms up, and pervades the house with the unmistakable aroma of cat spray. The cat will notice even quicker, and as the chemistry of the smell is probably altered a little by being heated the cat may think that that invisible rival has been in again and overmarked its mark.

Clearly most cats are not so easily disturbed by the arrival of new items in our homes or by the process of redecorating. They are often used to a constant barrage of novelty within our homes and after a simple sniff and exploration of the items or a quick tour of the newly arranged room are sufficiently reassured that all is as it was and home is still home. A few are more easily upset, particularly those that live permanently indoors and are unused to a frequently changing lifestyle full of daily challenges and decision making. For them new furniture or objects or even the arrival of visitors can be a major event, and perhaps sufficiently disrupting or threatening to send them over that emotional threshold of insecurity.

For such sensitive characters reducing the size of the territory to one unchanging and secure room or pen for a while may help the cat feel less emotionally upset. From there it can slowly be introduced to other rooms in the house, with the owners acting as security bridges at initial exposure. Rooms that have been altered or that are subject to continuing change, as, for example, when we build extensions to the kitchen or living room, should be available to the cat only when they are as complete and unchanging as possible. Such areas should be explored last of all following an outbreak of spraying. It should be said that living with such cats can mean that while life is calm there is no spraying, but any minor change in the home or routine could cause the problem to start again. So if you know change is coming, it may be wiser to board the cat until it's all over or

keep a supply of sedatives or other suitable treatments in the
cupboard so that treatment can be administered to the cat
and its reactions dulled a little before the onset of the
changes.

> *Dear Mr. Neville,*
>
> *I have owned three neutered cats, two girls and a
> boy, for over three years with never a problem.
> Two weeks ago a friend asked me to take on a
> homeless young kitten for a few days while she
> found it a home. Needless to say, I can't bring
> myself to part with the kitten now, even though
> she appears to have caused something of a dis-
> ruption for the others. George, the quietest char-
> acter of the original three, has begun to spray all
> over the house since Carla arrived. I would have
> thought he would have accepted a female kitten
> without any problems, especially being used to
> sharing his patch with two others already. What
> can I do?*
>
> *Yours sincerely,*
>
> *Janice Wooley*

If the single invasion of a rival through a cat flap, or even the
possibility of it, is sufficient to set some cats spraying, it is
not surprising that the arrival of a new cat permanently in
the house will set others off. While early experiences of
social living will usually improve any cat's ability to tolerate
others sharing the home later, there does seem to be an
individual limit for each cat. Perhaps it is due more to loss of
personal space than to competition for resources, but while
some cats like George start spraying as soon as the least
threatening form of feline challenge is presented, others live
quite happily in groups of fifty or more with not a sprayer to
be found. Indeed a wonderful character of my acquaintance
named Bernadette keeps more than 140 cats in a three-
bedroom house and an outside run in Harrow and without a

sprayer among them. More than this, Bernadette is constantly introducing new cats and temporary lodgers as part of her excellent cat rescue activities. The 142 residents (correct at time of going to press) are all neutered, exceptionally well cared for, and very harmonious in their relations with each other. It is quite a remarkable sight and perhaps leads one to believe that there is a threshold density of cats beyond which none will spray to try and maintain an impossible hope of personal space. This does not necessarily imply that they are suffering in any way as all Bernadette's cats eat well, play, and show no behavioral vices. Yet beneath this threshold it seems that the more cats one acquires, the more likely it is that at least one will spray. I can offer no guideline to owners of ten cats with a sprayer or two in their number as to how many more to take in until the compression threshold is likely to be reached!

With regard to sensitive cats like George, who are just below the threshold of spraying while living with two others, it is essential that any new cat be introduced very carefully indeed. A new kitten or cat should be housed in a pen for a while in the same manner as described in Chapter 4, for in this type of case we are considering a cat that cannot cope. Outwardly it may show no other signs of fear or aggression, but the spraying is a strong indication that the cat cannot maintain its personal space to the point of feeling confident. Slow introductions, with the new arrival confined to the pen and allowed only controlled access to the rest of the house one room at a time, will perhaps help George habituate to the new cat's presence, but more often than not he will need a little help from the vet for a few weeks. A low and decreasing dose of progestins helps cats like George to acclimatize slowly to the presence of a new cat and to develop a lower expectation of space and self-determination in the home. Treatment should be coupled with cleaning up and the use of food deterrents as outlined earlier.

Interestingly, neutered males seem most likely to be upset and spray at the arrival of a new cat in the house, especially if they already share with a female cat rather than other males. Perhaps the females are regarded as a resource and even the neutered male is simply defending or protecting his harem in such cases. Neutered males do, however, respond well to progestin treatment in most cases. Unneutered females may be most intolerant, perhaps because of the need for a larger personal area to allow for growth and development of kittens. Spayed females are more tolerant and seem to be more accepting than neutered males. But if they do react and start to spray, it is my experience and that of other behaviorists, that progestin treatment is not always effective at stopping the cat from spraying after the period of prescription. In short we can calm the cat's emotions so that it tolerates a new arrival and does not spray while the drugs are effective, but there is no learning or habituation process established to prevent the cat from spraying after the course of drugs is complete. In such cases it may be kinder and cleaner to put the new arrival in a new home as soon as possible.

> *Dear Mr. Neville,*
>
> *My Siamese cat Mao is a very loud, attention-demanding dog! When he is active, he expects the world to amuse him and for me and my family to act as his slaves. If he yells for food, we feed him. If he wants to go out, we get up and open the door, and if he wants affection we drop whatever we're doing to cuddle him. If we don't, he sprays us, so rather than be wet all day we pander to his every whim. It there a way out of this without upsetting Mao? He's even started to spray uncooperative visitors as well!*
>
> *Your obedient servant (of course),*
>
> *Margaret Thornley*

This is a typical example of the learned sprayer. Mao, being a brighter-than-average member of a brighter-than-average breed, has simply learned that he can obtain his owners' attention or anything else by spraying. If rubbing around the legs and purring didn't produce the desired results one day, his frustration or emotional upset at being ignored may have caused him to lift his tail. The owners, expecting the worst, immediately diverted all attention to Mao, and the first stage of the learning process was established. It sounds as if he has even learned to allow a short delay after yelling or posing for the owners to notice and respond appropriately and withholds the spraying if the favorable response is received. The tail is doubtless held up straight and quivering slightly at the tip to indicate the real consequence of failure to respond. Mao and his type will wait a short time, and if the desired results are not achieved the spraying ensures that at least some attention will head his way. It may arrive in the form of a shout and a size-twelve boot, but it's better than being ignored and is still a form of success to the demand. The owners, of course, progressively train themselves to respond ever faster to the cat's desires, and the cat becomes a focus of constant attention so that its wishes are not inadvertently ignored.

Treatment inevitably involves a little role reversal, with the owners reestablishing themselves as the instigators of relations with Mao and not the responders. This may mean a period of heightened spraying for a while, but it is essential that Mao's demands cease to be rewarded at any time. He has to be fed, fussed over, let out, and totally managed at the owner's whim, not his own, and none of his threats or actual spraying should ever be responded to at the time. Cleaning up takes place when he is elsewhere. Mao has to be ignored, though if he hasn't become less of a reactionary after a few days, use of water pistols or loud distracting noises as negative reactions to his threats can help recondition him. Such tactics have to be seen as remote punishments, however, and not as punishments from the owners. Otherwise this too may be seen by Mao as success in his effort to procure attention. Such conditioning can work only with the more intelligent cats that have already learned such sequences of behavior. In essence cats like Mao are being treated much more like dogs, and the techniques of restructuring learned or uncontrolled behavior will be much the same in both species.

Usually only a short period of restructuring is necessary, and it is simply a case of making one's mind up to do it and living through the consequences of owning a confused cat for a few days.

The bad news with such bright cats, who are clearly experts on how to shape human behaviors, is that they learn new ways of demanding attention without our realizing. The cat walking along the windowsill sees how quickly we leap to our feet to pick him up as he brushes past the family heirloom vase full of flowers and simply heads to exactly the same spot the next time he wants attention. Alternatively he may learn that chewing on a favorite plant or scratching the furniture is just as effective as threatening to spray. Like living with highly intelligent or hyperactive children, life with a bright cat can simply become a case of trying to keep pace with the latest trick.

> *Dear Mr. Neville,*
>
> *We've almost grown used to our Burmese cat, Baroness, spraying when she's upset or excited or when we bring anything new into the house. We've always tried to be careful and give her plenty of opportunity and help to explore new things. If we have a bad period with her, we confine her to the kitchen for a while and then allow her into the rest of house again gradually, and this seems to work better than using drugs to keep her calmer. But last week she expanded her repertoire and included my fish and chips! They aren't supposed to spray food, are they?*
>
> *Yours sincerely,*
>
> *Helen Drysdale*

No, Helen, they aren't supposed to spray food. Perhaps your fish and chips was of such poor quality that Baroness decided it needed a little extra vinegar before she could possibly consider eating any. I'd have a word with the fish and chip shop about the quality of its food, which clearly isn't recognized as such by the cat, unless she was objecting to the quality of newspaper it was wrapped in.

A flippant reply to cover my total disbelief that a cat should do such a thing and to cover up a feeling of helplessness about what to do. Baroness is recorded here simply to illustrate that yet again there are situations when cats defy explanation and treatment. We think we understand the mechanisms and motivations of some of their behavior, and experience in dealing with problems leads to the development of ideas that can help with treatment. And then along comes a cat like Baroness and sprays what one has come to regard as a form of novel intrusion that cats will tolerate and one that might even be a deterrent to spraying. It only serves to keep one on one's toes and reminds one never to take anything for granted with a cat. As for Helen, she'll have to try the local Chinese take-out. Such offerings are usually irresistible as food, not spray posts, to all cats, oriental or not.

Middening

In case you are now tempted to put his book down to go and eat something, we will move on to another form of marking behavior in cats that is even more distasteful than spraying, that of middening.

Dear Mr. Neville,

Since we moved a few weeks ago, my cat, Snaggle, has persistently poohed both by the back door and the hallway by the stairs. I've tried offering litter trays—in fact there are now six throughout the house—but if I put one where Snaggle has soiled he simply moves a little distance away. I have only caught him at it once,

and when he saw me he behaved perfectly nor-
mally, as if he had done nothing wrong at all.
What do I do to stop this behavior?

Yours sincerely,

Anne McLeish

Middening is not a function of misplaced or misdirected toileting. It is a very definite act, using feces placed in the open at specifically chosen sites to identify the perpetrator's occupancy or right to use a certain area or walkway. It seems most likely to be performed by pet cats that are already a little nervous or incompetent in their general demeanor and are then subjected to some form of major upheaval or challenge. The same types of upset that may cause one cat to spray may cause middening in another. Worse still, the behavior is often more difficult to resolve than spraying because it represents a larger-scale reaction to the challenge and indicates that the cat is more emotionally disturbed. The act of middening is usually carried out by doors or on walkways used by the family, as it is here that strange, unfamiliar, or challenging smells from the competitive areas outdoors are brought right into the core of the cat's territory. But other cats choose even more public areas than this to midden. Some try to establish rights-of-way on top of chairs or refrigerators and in their beds. Others midden in more personal items of ours such as stereo headphones. Electrical equipment such as video machines and televisions seem especially popular with the middening cat, and one acrobat I once saw even managed to midden on top of a door that it used to pause on midway between jumping from one cabinet to another!

Many species of wild cat, including tigers, do not always cover their feces and seem to leave them in the open as a deliberate mark. Similar behavior is observed in other animals such as otters and mink, which leave their fecal marks, known as *spraints*, on rocks and high points along riverbanks to help define their territories. Middening is more often observed in feral cats living in high densities in cities than in farm cats with more freedom to move on if stressed or threatened by other cats. Dominant cats in any environment do seem to midden more often on high relevant points within their territory than subordinate or less assertive individuals, and unneutered pet males are reported to midden in areas under dispute with other local house cats. Middens are perhaps best interpreted as unavoidable blockbuster visual and odor messages, designed to drive the point home, as compared with the comparatively subtle and longer-lasting, more complex patterns of odor left in squirts of sprayed urine. Pet cats that midden must be treated delicately along the same lines as sprayers and not as dirty or house-soiling cats. Close confinement in a secure pen for a week or so is often the first step, with gradual exposure to as much of the rest of the house as the cat can feel confident in. Sometimes this will be only a couple of rooms, but if the cat seems happy with that much I tend to advise that further progression not be forced on it. Following identifiable disturbances such as Snaggle's move, controlled exposure to new rooms, neighbors, and the yard is always to be advised and may be assisted by a little sedative treatment for a few days. It will also help if owners remember not to bring in too many items from outside for a few days and to remove their shoes outside the door and put them in a closet before entering until the cat gets used to the new range of smells in its new outdoor environment.

I get to see very few middening cats, and this perhaps indicates how rare the behavior is indoors, though I suspect a great many of our pet cats, particularly in high-density feline suburban populations, midden outdoors at disputed areas of territory or when arguments arise over which cat has access to certain points at what time. Such middening is most likely to occur away from the owner's property, often on a nearby neighbor's shed or garage roof, yard paths, or feline walkways along the base of our yard fences. I also suspect that if cats start to midden indoors, owner tolerance and understanding levels are not likely to be high, and if the cat doesn't respond to the provision of extra litter trays (and I wouldn't expect it to), then it will not remain an indoor pet for long. If it is lucky, it may get to be an outdoor cat with the same owner or it may go to a rescue home. But sadly it may pay the ultimate price for being overresponsive to change or challenge.

৯ 9 ৫

Indoor Toileting

There is no animal quite so fastidious about its personal hygiene as the cat. Its meticulous attention to toileting and maintaining a clean home base is one of the main reasons for the cat's popularity. While some dogs and even rabbits can be trained to use a litter box, no other domestic pet will dig a little hole to squat over, "do its business," and then cover it up to avoid causing offense. Whether the cat uses the yard outdoors or a litter box inside, the vast majority of cats never make the mistake of leaving unwanted presents, except perhaps when ill or when we accidentally lock them in. The cat that is forced to toilet where it would otherwise not usually bears an extremely unhappy disposition for some time afterward, so strong is the desire not to soil its home or sleeping area.

How should this behavior arise, and why is it that while young puppies may require some weeks of careful attention to become housetrained, kittens arrive at six or twelve weeks of age already set in their habits? We can only guess that there is an evolutionary advantage, especially to the young, gained by not soiling near the resting area to avoid attracting the attention of predators or prey and to help ensure a food supply nearby. It is also suggested by American behaviorists that burying feces helps disrupt the life cycle of some enteric worms and so reduces the risk of infection to the cat and especially to vulnerable kittens. I am unconvinced by this suggestion because the tapeworm needs to cycle through a rodent's muscular tissues and can reinfect another cat only if it eats the infected rodent. Perhaps a theory gone too far, though I have little to offer in its place except that it seems to make sense that a cat, like a dog or a person, would be most unwilling to soil in or near its own bed because of the unpleasantness of lying close to waste or in the damp. Those animals prepared to do so would also place themselves at greater risk of infection from bacteria, fungi, etc., and so be at a disadvantage to those living in a dry, clean bed. As a result they would be less likely to survive infancy to pass on their lack of hygiene to the next generation.

The cat's attention to hygiene in the nest is predetermined at birth. The young kitten is physically unable to urinate or defecate without stimulation of the abdomen and perineal (bottom) region by its mother's licking. This is known as the *urogenital reflex* and results in kittens' passing waste only when the mother is present to clean them up immediately. Therefore the nest is kept clean and the release of odors that might attract predators when she is unable to protect them is avoided. The reflex can remain operational until the kittens are about five weeks old, though most can urinate and defecate voluntarily by three weeks. As anyone who has had to hand-raise kittens is aware, physical stimulation similar to the mother's must be carried out to ensure that the kitten excretes. This is usually carried out after feeding, though as the kittens grow up approaches can be made at any time and will cause the kitten to adopt a passive position on its back with legs open to receive the necessary attentions. Older kittens are more active, and their own play and wanderings will serve to provide forms of stimulation and elicit toileting.

As the kittens start to wander and fall out of the nest to explore, their mother will continue to look after their toileting needs and they will demonstrate the urogenital reflex outside the nest. Indeed she may even carry them away from the nest for this purpose. This not only establishes the idea that the kittens are unable to toilet in the nest for hygiene reasons but also effectively trains them for life at the most impressionable of ages not to want to soil the nest. This early-established antipathy helps ensure that subsequent generations will be taught similarly, and it is also an extremely useful feature to employ in the treatment of toileting problems when they do occur.

Outside the nest all kittens have a natural tendency to paw and rake at loose earth or soft, movable material. Observing the mother's toileting behavior and specifically her efforts to excavate a latrine hole and then cover it up after use helps the kitten adopt similar toileting behavior. Also, the smell acts as an attracting focus when nature calls, and it directs the kittens into using particular latrines. By the time they are weaned at about six weeks old they are already using soft soil for their toileting functions. In short kittens housetrain themselves for the most part. All we have to do when allowing our cat to have kittens is to provide a suitable litter tray for the kittens to explore and learn to associate as their latrine in those first crucial weeks. The association with litter usually transfers well to soil if the cat is later allowed outdoors, thus helping the cat perceive our home as its sleeping/feeding nontoileting area. For cats that are to remain indoors, toileting can be transferred to virtually any of the huge number of different litter products available for our cats while they are young, though most have their preferences and may be less willing to accept change later, when adult.

The earth-raking behavior of cats may occur in simple response to the presence of feces and accounts for the enthusiastic covering by some cats long after the feces have been obscured from view. They can still smell it, so the raking behavior persists. Others rake over the area nearby or, as is observed in many pets, the area alongside the tray and other objects within range even though they are different in texture and appearance to the litter substrate. For some the need to cover is very strong indeed, and they cover not only their own feces but also those of other cats. Observing cats at their toilet may not be the most rewarding or edifying of pastimes, but it is interesting to observe that the cat will often pause between rakes to investigate the smell of the covered feces. Raking over continues until the cat is happy that the smell has reached the right level. This may mean total disguise so as to avoid drawing attention to the depositor, or it may mean that the smell persists through the soil as a deliberate gesture to inform other cats of its presence. At this point what appears to be toileting may have more of a middening function. Rather than toileting in the open, whereupon the feces may weather very quickly, it may help to provide a fresher and more persistent mark. Again, our sense of smell lets us down. Even though the smell of cat feces is highly volatile and detectable, we are unable to decipher that delicate package of odor information so as to learn exactly what is being said.

By and large cats are well suited to us because of the ease and lack of effort needed on our part to cater to their toileting habits. Either they are perfectly happy and conditioned to use a litter tray and will strive to get there even when seriously ill, or they will use the yard. As our neighbors' yards are farther from the center of our cat's world than our own, the cat may prefer to go there. However, this may occasionally cause some dispute, especially if your neighbor has recently tilled his vegetable patch and produced that soft topsoil that is the favorite latrine of many cats. Others deliberately leave their feces uncovered or only partially covered as a mark in their territory under challenge from other cats, and this too is perhaps more likely to occur away from their own home. But as our cat's territory of home and yard is also the challenged edge of another local cat's territory, we are probably just as likely to receive as our cat is to give, and between cat owners, at least, all is usually about fair. Non-cat owners may be unconvinced about the beneficial acts of fertilization of their garden so freely provided by our cats but, if seriously disturbed, can always acquire a Jack Russell terrier to ensure that no cat ever feels confident enough to adopt that most vulnerable of postures in their garden.

The upset experienced by the owner of a kitten that has never become so well housetrained or, perhaps worse, of the cat that has suffered a breakdown in previously immaculate toileting habits can be severe. One of the fundamental reasons for choosing a cat as a pet is totally undermined when it leaves that unmistakable package in the middle of the carpet or behind the chair or, perhaps worse, a wet patch on the bed or morning newspaper. While most of us will accept an occasional accident or a temporary loss of house-training due to illness, persistent problems are sufficiently disturbing to account for about one-quarter of all the cases referred to me. This contrasts with the very small number of canine toileting problems I treat. Perhaps dogs are even cleaner than the fastidious cat, or is it just that we expect it more from a dog or blame ourselves because we are ordinarily responsible for ensuring that it goes out? Perhaps dog feces are simply less offensive, more openly deposited, or shorter-lasting in their offensive odor than cat feces.

Medical Problems and Their Aftermath

Dear Mr. Neville,

Since my cat Marcus suffered from cystitis some weeks ago, he has consistently urinated around the house. Solids are still well on target in the litter box, but he now seems to urinate wherever he chooses. The cystitis has been cured by my vet, so there is no medical reason for this to continue. Do you have any ideas as to why this should happen, and is there anything I can do to house-train Marcus again?

Yours sincerely,

Leonorah Jones

Medical illness, especially urinary cystitis and mild arthritis, which occurs most frequently in the older cat, are common causes for a breakdown in indoor toileting habits. While solids continue to be passed in the tray or outdoors, urine needs to be passed more frequently, so the cat may get "caught short" more often. With general old age the cat's urinary sphincter may loosen, making retention of urine more difficult, and arthritis may also make retention more painful, so the cat will need to urinate more frequently to relieve the discomfort. Such conditions may force the cat to urinate in places other than its usual litter tray simply because it can't get there in time or it is too painful to do so. Some cats understandably associate the burning pain of urination when suffering from cystitis with the litter tray and avoid it deliberately, selecting other, "less dangerous" sites. Once new areas are used as latrines, they may continue as such long after the medical condition and any pain have been relieved or cured. It is almost as though a new association to the different substrates of carpets or soft materials becomes established as equally acceptable to the cat as the litter. While many cases do resolve in time, most can be usefully assisted by providing the same association opportunities as the cat experienced as a kitten.

During treatment of the medical problems it is often a good idea to house the cat in a much more restricted area than usual. One room, preferably a small one such as the bathroom, will ensure that he or she is never more than a few paces from the litter tray. If no small rooms are suitable, then a greater number of trays should be provided. The aim is for the cat to travel as short a distance as possible should nature call and to maintain an association with the litter as its latrine throughout treatment, thus precluding or reducing to a minimum the chances of urinating elsewhere. Each time the cat needs to urinate, it will be close to or on the litter tray and so, by a process known as *successive approximation*, will always get it in the right place through lack of opportunity to get it wrong. After a few days of consistent aim the cat can be allowed more freedom, with extra litter trays available in each room, until the medical aspects of the problem are resolved. The number of trays can be reduced steadily thereafter until the original number remains. However, older cats may always need closer confinement when unsupervised, or more trays around the house if particularly stiff and arthritic.

Similar latrine association problems have been encountered, though more rarely, with cats suffering from diarrhea. This can occur as a result of gastric upset or as a secondary problem from upper respiratory tract infection, which may cause excessive mucus intake and lead to diarrhea. Similar treatment can be applied with usually successful results.

Dear Mr. Neville,

When my cat was ill, my vet prescribed pills to be given three times daily. Fantasia, my cat, soon learned to keep out of the way when the time came for her to have her pill, and so I had to take advantage of her when she appeared. This was most successful when she was on her litter tray. We finished the pills and, thanks to my vet, she is now recovered, except for the fact that she will no longer use her tray consistently for toileting. Instead she has started to use other areas, in the corner of the living room and behind a chair in my bedroom. I feel we are "out of the frying pan and into the fire." What should I do to convince Fantasia that there are no more pills to be given and to make her use the tray again?

Yours sincerely,

Serena Cuthbertson

Oh dear! It's never a wise move to take advantage of anybody when at his or her most vulnerable, no matter how much easier it may make life at the time. Fantasia now clearly associates the litter tray as a dangerous place where unpleasant things may happen. The main thrust of treating this type of self-inflicted problem is for owners to make the cat's litter tray as secure as possible. In the same way as Fantasia is now selecting secure corners or areas behind chairs to give solid protection from interference, the owners should move the litter tray to just such a corner or, better still, to a corner and under a table to preclude "attack" from above. The move itself will help reduce the cat's association of ideas. Still more important is to provide a covered litter tray by purchasing one of the proprietary brands or simply inverting a cardboard box over the tray but inside the lip to ensure that the litter stays in the box. A hole the size of the cat flap should be cut in one end for access, and once inside the cat should feel safe. This simulates the type of area selected for toileting by many cats outdoors, such as under a low bush next to a wall. Many cats are more confident and will toilet confidently in the open, especially on freshly raked soil, but, like the proverbial ostrich, will turn their backs to any on-lookers as they squat and perhaps pretend that the onlookers are not there. Needless to say, to avoid such problems as Fantasia's a cat should never be disturbed for any reason while toileting on a litter tray, and should only be bothered outdoors if soiling where you are about to plant your vegetables.

Early Learning

Dear Mr. Neville,

I have recently taken on two three-week-old abandoned feral kittens and am having to hand-raise them. While I dutifully tickle their stomachs and rear ends after feeding, I have yet to notice any action on their part to dig into the litter that I place them on. Is it possible to teach them to dig, or will they do it in time anyway?

Yours sincerely,

Linda Court

One of the problems of hand-raising is that while we can now provide more than adequately for the kitten's nutritional needs, we still rely on instinctive behavior patterns to develop without any help. Normally they do, and one would expect such young kittens to show the litter-moving and -digging reactions soon. However, we can never fully substitute for the mother's teaching or for the opportunity for the kittens to observe and learn from their mother's behavior. With this type of problem it will help to manipulate the kittens' paws gently in the litter after their feeds, and hopefully they will soon pick up the idea. It may establish itself more fully later when the kitten observes another cat digging a latrine hole. The important thing is to provide plenty of opportunity for the kittens to play in litter and to develop their interests in making it move, as a precursor to digging little holes.

Lack of early opportunity to experience soft litter or soil can result in the city feral kitten never associating it as its latrine. Even if successfully tamed later, some may experience difficulties in learning to use litter for toileting, having grown up on concrete or hard-packed soil. Open toileting for urine and feces becomes established and may be difficult to redirect to what we would prefer the cat to use as a latrine. Fortunately most cases can be resolved through the application of fine litters in a confined area for a while. Most seem unable to resist the opportunity to dig into fine, dry sand or perhaps one of the new generation of fine, light particle litters such as Ever Clean. Given no other alternatives for a few days, the behavior soon switches.

Older cats given the luxury of a home later in life, perhaps having strayed away or been rejected from it earlier, may have been forced to toilet on solid surfaces during their spell of life on the wild side. They too may need a little time and forced opportunity to readapt to using litter. With them it may help to provide soil in the tray initially but thereafter to change gradually to a preferred cleaner commercial litter. For feral cats that are adult when captured and tamed (ideally they should be neutered, treated, and returned to the site where they were captured whenever possible), adapting to litter tray usage may be a very slow process in a few but is usually surprisingly quick. Often the tamers erroneously interpret the use of the litter tray by the feral cat as the first step in its domestication. It isn't. The cat is simply doing what comes naturally.

A word of caution for all inexperienced cat breeders: never be tempted to bring in soil from the garden as litter for young, unvaccinated kittens. Soil may carry many infectious agents that can cause serious illness or even death. However, the smell of soil can be applied to an ordinary cat litter and so ease the transition from litter to soil when the kitten has been fully vaccinated and can be safely allowed outdoors, usually at twelve to fourteen weeks of age.

The smell of a cat's latrine is a key part in ensuring accurate usage. Kittens that later miss the tray or have trouble in associating it as their latrine can often be encouraged to use it by placing the "mistake" in the tray rather than disposing of it. The kitten will be more likely to use the tray the next time and may even feel the need to rake over the previously misplaced effort.

Food Proximity

Dear Mr. Neville,

When we obtained our kitten, we were delighted with how clean he was. We have lavished attention on him, bought him all manner of toys, and established his own little corner in the kitchen for bed, food, and litter tray. After a few weeks he stopped using his tray and began soiling the mat by the back door. He continues to sleep in his bed and eat his food, but will use his tray only if we put it on the mat or away from his corner. We have only a small kitchen, and this is rather inconvenient. Is there anything else we might do?

Yours sincerely,

Sheila and Michael Pryce-Jones

Isn't it strange how many people assume that cats are so immaculate in their toileting habits that they would even be prepared to eat near their toilet? We certainly wouldn't dream of taking our dinner into the loo, yet in the urge to compartmentalize our pet's needs we may expect exactly this from the cleanest and most sensitive of our pets. In the same way as food can act as a deterrent to soiling and spraying in the home, it can also deter cats from using their preferred toilet, the litter tray, if it is placed too near. Food must always be placed as far away as possible from trays so that the cat is not forced to use other, undesirable latrines. This is one of the most common reasons behind the onset of house soiling by cats and definitely the most avoidable.

Latrine Association

Dear Mr. Neville,

My cat Fred developed quite normally as a kitten and was most clean in his toileting. However, at about four months of age he suddenly began to use the carpet in three or four places around the house and now refuses to use the litter tray for either solids or liquids. We have at least two trays in every room, but he ignores the lot and simply heads to one of his places. If we move a tray there, he simply moves to a fresh spot. What should we do?

Yours sincerely,

Jodi Burton

Just occasionally a cat may steadily decide that the litter provided or the tray and its position are not acceptable as a latrine area, even when food is nowhere near the tray. Sometimes this may arise because of more general nervousness or simply because the cat prefers to use the carpet. Once the new substrate is established, it is very unlikely, if not impossible, that the cat would spontaneously revert to using litter again. Instead opportunity to soil the carpet is reduced so he will develop a preference for another substrate. This is achieved by confinement for a short period in a carpet-free room, such as the bathroom or kitchen, or in a pen with a wooden floor. A litter tray is offered, lined with carpet. Usually the cat is impressed by this gesture, if not the confinement, and readily uses the box. This reestablishes the idea that toileting occurs in that place. The carpet is then speckled with a covering of lightweight litter or an alternative such as wood shavings or unprinted paper shreds. The layer is made thicker at every cleaning, and after a few days of good, consistent use the underlying carpet is entirely removed. The cat is then allowed progressively more freedom in the house and will usually stay bonded to the idea that toileting indoors occurs on a given surface, in a particular place, and in a certain receptacle. For best results the cat should have a period of at least two weeks before any contact with carpets is allowed without supervision.

Cleaning of unwanted toileting areas is as essential as cleaning up indoor spraying posts. All soiled areas should be cleaned with enzymatic liquid laundry detergent and then sprayed with alcohol. These should be allowed to dry thoroughly before the cat is allowed any access at all, supervised or not. Certainly no ammonia-based cleaners should be used, and preferably none that contain chlorine such as household bleaches. Bleach and the phenolic compounds found in many disinfectants are very toxic to cats, and their use must be avoided wherever the cat may roam, especially at his litter tray. The smell of a latrine is probably as important as the substrate and position of it and serves to attract the cat and build up the association of that place as a toilet. Thorough cleaning to remove all traces of that smell is therefore essential, and unfortunately some deeply soiled carpets or furniture may have to be discarded to achieve this aim and for treatment to be successful.

Again, it is my experience that none of the standard proprietary repellents to indoor soiling are effective, nor are any of the traditional methods such as applying ground pepper to the soiled area. Food does, however, act as something of a deterrent and can usefully be placed at unwanted latrine sites once they have been cleaned (see page 140).

> *Dear Mr. Neville,*
>
> *For over eight months we have tried to get our beautiful Persian kitten, Khan, to be clean indoors, but he just doesn't seem to realize what the litter tray is for. We have tried all types of tray, litter, and places, but to no avail. He simply goes where and when he likes, even in front of our very eyes. He really has no idea of what to do, yet in other ways is normal and very affectionate. We have tried being kind, and we have tried punishing him though we know it probably won't help. Please, can you help or we may have to put him to sleep or keep him in a cage all of his life.*
>
> *Yours sincerely,*
>
> *Anna Selwood*

The desperation of trying to treat a kitten that just will not show any associative behavior in its toileting activities can be enormously frustrating, as Khan is proving to his owners. For some reason, doubtless as a consequence of being highly bred, the Persian is the greatest exponent of all the breeds at failing to learn. Well over half of my house-soiling cases concern this breed, which is a pity, because they are often too expensive to carry such a burden. They are also too beautiful, if you like that sort of thing, to be dirty in the house, and owners do indeed become exasperated in trying to get them to be clean. As Persians demand much attention with regard to grooming and maintenance, owners perhaps feel more cheated than most, and as few Persians are intended to live outdoors, the option of putting them there for most of the day cannot be considered.

With extremely serious cases of breakdowns in toileting behavior, or with kittens, there is really only one course of action to follow. It isn't always successful, as the presence of other cats may be influencing the sufferer's toileting habits, but it has put a few diehards on the road to indoor cleanliness. The cat is confined as before in a pen, but with only a bed and a small area of floor space. It is brought out to feed at mealtimes and more frequently given access to water. The whole of the floor area is then covered with litter, preferably a fine, light brand. The cat will usually have developed the idea to move away from its bed to toilet and so will do just that in the pen. If it does soil the bed, then this is removed and replaced with a cat-size square of carpet. Then the cat's only option is to soil on the litter-covered floor, and gradually the size of this floor space is reduced. Progressively over about two weeks the area is brought down to the size of a litter tray. A plastic, sideless sheet is then placed under the litter, and then later a small-sided tray is used to house the litter. Usually the cat continues to aim for the litter throughout the period and gradually accepts a smaller and smaller area of litter surface as its latrine. The space where the litter used to be on the floor of the pen can also be occupied with tubs of dry food in the later stages of approximation to continue to force the cat onto the litter, again working on the premise that the cat will not voluntarily soil on or near its food. The only drawback with all this is that Persian cats will have to be cleaned and groomed far more frequently than usual, especially in the early stages of confinement. But if all goes well, the attachment to litter is usually as strong as with any other self-taught cat, and the patient can be allowed out of the pen into one room and then gradually into others.

Why Persians? There may be some inbred inability to show interest in loose litter early on, or perhaps they are simply slower or unable to associate their mother's latrine behavior as being something to emulate. More likely is that the behavior is undeveloped in the mother and the kitten has no opportunity to learn by observation. Poor mothering ability in any cat, pedigree or moggy, can cause all types of problems in kittens from inappropriate toileting behavior right through to death from neglect. But poor mothering ability is not the reason behind every case, as perfectly house-trained Persian mothers may also produce an occasional kitten that doesn't follow her good example. So really we need special efforts to be made by breeders of Persian cats to identify such kittens as early as possible so that they may be given special attention, with paw guiding for digging and investigatory play in litter. Perhaps then I'll see a few less, though to be fair to Persians generally I don't see that many. It's just that they always make the most difficult patients when it comes to house-training.

Anna Selwood had been moved to punish Khan for his toileting errors indoors as an act of desperation and much against her better judgment. All punishment is pointless in trying to treat the indoor toileter, especially the old nonsense about rubbing the cat's nose in it. The poor cat will have no idea what is happening after the event other than the fact that its normally friendly, motherlike owners are behaving in a frightening and unpredictable manner. The same will be true even for the cat caught in the act, for whom there is no crime other than wanting to relieve itself. Punishment will not help redirect the toileting behavior into the waiting litter tray or make the cat anything other than nervous about the approach of its owners. And a nervous cat has less control over its urinary and anal sphincters. Patient application of a few simple techniques is the only way to treat the cat that soils indoors and never the attempted use of negative conditioning.

Cleaning

Dear Mr. Neville,

When we had only Morris, there was never a wet patch in the house, other than in the litter tray. Then we decided to get Morris a little friend, Marla, and ever since there have been occasional wet patches made on the kitchen carpet near the tray. Should we clean the tray more frequently now that "we are two" or provide extra trays?

Yours sincerely,

Brigette and Olivia Simons (ages 9 and 11)

Cats do not like to use dirty litter trays, and doubling the
input by doubling the number of cats is a sure sign to double
the number of trays or, better, to double the frequency of
cleaning. The main reluctance probably stems from the cat's
dislike of standing on damp surfaces and the smell of litter
that has been overused. It may help to change the litter
gradually to a more absorbent variety that clumps when wet
and so concentrates the smell and damp. However, there is
little substitute for regular cleaning. Some cats do not like to
share trays even with friends, and there is little option here
other than to provide more trays or clean out existing ones
after every usage. Again the clumping litters may mean less
work for the reluctant cleaner as only the clumps and feces
need be removed for a while, and complete changes of litter
will be required less frequently.

Substrate Preference

Dear Mr. Neville,

*We used to use a gray earth type of cat litter for
our cat but found this very awkward to carry up
the stairs at our new apartment. It was even
heavier coming down once used, so we changed
to using a scented wood pellet variety. Since then
Tom has been most reluctant to use his tray, and
we have now had to go back to using the gray
earth. I would have thought the pellets would
have been more hygienic for him and would very
much like to use them. Is there anything we can
do?*

Yours sincerely,

Pauline and Bertie O'Sullivan

Cats can bond very specifically to one type of litter, and a
sudden switch from gray earth granules to wooden chip
pellets, with added pine or other scent, is a far greater
change than we realize for the cat. Many are unimpressed by

the advantages of such a modern lightweight approach, and others are seriously deterred by the scents that are added to some litters to please the human purchaser. This seems especially true of those that release the bulk of their scent as they become wet from the cat's urine. The whiff of fragrant pine may deter the cat from raking and perhaps from using the tray altogether. Discouraged and not a little confused, the cat will often appear desperate to seek other, less disturbing places than its tray, thus quickly building up an association with the carpet or elsewhere as its latrine. So choose your cat's litter with care and never introduce changes suddenly, as these can cause a breakdown in toileting habits very quickly.

Where a change is to be made, the old litter should be phased out gradually in favor of the new over the course of about a week so that the cat slowly acquires the new association of texture and smell. A small amount of used damp litter from the previous day should be added to the new mixture to encourage smell association. This technique can also help cats start using a tray again where there has been some breakdown, as it helps them recognize the latrine for what it is. The trays of cats with house-soiling problems should never be cleaned out too often as this precludes the development of the latrine association. Once every two days per cat—that is, once a day for two cats—is about right for the average-size tray, and this can be increased to once per day per cat once the problems of soiling elsewhere are resolved. This is a little less hygienic than could ordinarily be recommended, but provided the usual precautions of washing the trays and wearing rubber gloves are observed, the risk to the owner is not appreciably greater, and the treatment will help. Of course the cat must be up to date with antiworming treatment at all times.

Dear Mr. Neville,

Our cat Roger is not the brightest of cats but gets through life with cowardly bluff. That is, until he misses his tray. He knows enough to go to the

*tray but seems unable to aim things in the right
direction. Most times he climbs in the tray and
then hangs his bottom over the side to deposit his
offerings on the floor. If we put paper under the
tray, he doesn't get in the tray at all and simply
squats by the side of it. Is Roger stupid, or are we
doing something wrong? Please help.*

Yours sincerely,

Mary and Alan White

Roger has at least got some things right. He clearly knows
that the tray area is his latrine, but not how to position
himself or how to deposit urine or feces into the litter itself.
It seem that Roger doesn't dig a hole either, so perhaps his
early learning was incomplete or was interrupted at a crucial
stage. While some cats can simply be treated by offering a
large tray, or one with higher sides, the usual form of treatment is once again to remove the opportunity for error by
confinement and by covering the tray to ensure that he
cannot hang his bottom off target. Many cats whose attachment to litter trays is incomplete are reluctant to enter the
confines of a covered litter tray as they do not see it as the
secure area that most cats do. For them it is perhaps a
restriction that builds on an established antipathy toward
using the tray. Therefore a roofless cover helps direct the aim
without the cat's feeling restricted, and, if coupled with a
preferred finer grain litter, Roger and his affable type
shouldn't continue to make mistakes. If possible, such cats
should also be shown how to dig by manipulating their paws
if they allow it. Where such an effort frightens the cat, the
movement should be discontinued immediately so as to
prevent the cat from developing a reluctance to go anywhere
near the tray.

Dear Mr. Neville,

*Puss, our tabby female cat, will often urinate in
the bath or kitchen sink rather than in her tray*

or outdoors. Is there any reason for this? It isn't a
great problem as washing it away is easy,
though we do have to remember to do so before
washing up or having a bath.

Yours sincerely,

Stella and Norman Smith

This is a habit of quite a large number of cats but is usually
problematic only if solids are deposited. There may be some
latrine-type smells emanating from the outside drain that
percolate up through the plug to attract the cat, or it may
simply be that the sink or bath is one of those nice secure
toilets protected on all sides and therefore more attractive
than open trays. It is always worthwhile looking closely at
how the cat uses existing facilities and making them more
attractive by moving them, covering them, or changing the
litter. Thereafter simply filling the bath or sink with a few
inches of water when not in use will deter most cats, though
if this can be coupled with offering a normal tray in the same
room, the cat will hopefully divert to it when finding the bath
or sink latrine unacceptable. One can go as far as sprinkling
litter in the bath and then in a litter tray in the bath. The
tray is steadily raised up and out of the bath over a period of
a few days, moved over onto a chair, and finally onto the
floor and out of the bathroom. This seems a bit long-winded
but will usually work. It can't be recommended in the
kitchen for hygiene reasons, so the water technique is better
all around.

Nervous Urination

Dear Mr. Neville,

Mimi is such a timid cat. She seems frightened of
every loud noise or sudden movement and runs
away immediately when visitors arrive. While
she normally is very clean, after any disturbance
she may urinate in a corner or under the bed or

sideboard. On one occasion when a relation brought his dog when visiting, she even defecated under the bed in fear. Is there anything we can do to make her feel better and our flat cleaner?

Yours sincerely,

Miriam and Charles Bewley

Nervous urination is one of the most common forms of indoor soiling that I see, but nervous defecation is fortunately a rare entry in the diary. I well remember the rage of one owner whose cat had managed to leave a smelly pile in his hi-fi headphones, but as a new home was found for the cat posthaste I was unable to effect treatment. If the owner's temper and lack of humor were indicative of a general intolerance, then I suspect that treatment of what was a very nervous cat would have proved most difficult anyway.

Treatment of nervous urinators and defecators involves the application of the suggestions in Chapter 4 to raise the cat's competence at coping with challenges, such as the arrival of visitors. In severe cases this will involve controlled exposure to such challenges by penning the cat, so a litter tray should be provided to force the cat to use it in these situations. Raising the animal's competence is the main thrust of treatment, but support drug therapy with progestins is not recommended for nervous toileters. Progestins can cause water retention as a side effect and make the cat need to urinate more suddenly and at higher volume. This would be counterproductive to treatment, so other forms of mild sedation, such as Valium or some homeopathic remedies, should be considered first in such cases.

The cleaning of areas soiled by the nervous urinator is essential to prevent them from becoming established as latrines, but as most nervous eliminators urinate or defecate involuntarily few sites usually will be used more than once.

Some nervous urinators do so only when chastised by their owners, and by contrast these cats may head for a sheltered area to hide and then urinate or defecate. Never was the point so well made about being consistent with a cat and avoiding punishment for misdemeanors. Owners must always be pleasant, because for some sensitive individuals their whole security base collapses when their owner directs aggression or threats at them. Most cats are not quite so sensitive, of course, and may need an occasional reprimand when attempting to climb the curtains or walk across the table during Sunday lunch. The important thing is to make such intervention more startling than punishing and as untraceable to the owner as possible. A loud hiss by the owner will be interpreted as a general warning/threat in feline terms and inform the offender to escape from the area it is in. However, use such methods with caution with nervous cats, especially those that urinate when disturbed.

Associative Elimination

Dear Mr. Neville,

Last weekend my wife and I went away for a break and left our cats, Tom and Jerry, the run of the house rather than board them. Our neighbor came in twice a day to feed them, clean their trays, and see that all was well. They continued to eat well and seemed undisturbed by our absence with a very unpleasant exception. They used our bed, favorite armchairs, and a few clothes we left on the bedroom floor as toilets. Solids and liquids awaited our neighbor every morning, and naturally she is now unwilling to look after our cats when we go away in the future. What should prompt a cat to do such a foul thing, and more importantly, how should we prevent it in the future?

Yours sincerely,

Gwen and Harry Clarke

This type of indoor soiling is sometimes classed as "resent-ment" toileting but probably rather unfairly as, although it seems a deliberate act of vengeance or revenge, it is probably more the action of a disturbed or very nervous cat. Some go as far as urinating on the owner's lap, and a few more will occasionally urinate or defecate on the bed while the owners are in it. It is very hard to feel sorry for such cats at such times, but avoiding punishment is essential. The suggestion offered by the editor of this book that owners should take revenge on such cats and pee in their bed in return is also not recommended! Short-term fears or frustrations such as the withdrawal of a protecting influence when leaving the home for a few days or even the arrival of a new baby can cause the onset of such behavior. However, it is essential to review the position and acceptability of the litter trays or access to the outdoors for toileting and to ensure that these are adequate for the cat. Assuming that the cat has perfectly reasonable facilities and is not simply wandering in search of a toilet, we can usually assume that urinating on piles of dirty clothes, the bed, or favorite armchairs occurs when the owner is absent, and it can be interpreted in one of two ways. The first is that the anxious cat is simply sleeping or resting where it normally sleeps but on waking feels nervous or isolated and involuntarily urinates or defecates. Having done so once, the cat then sees the area as a latrine rather than a bed and continues to use it as such, seeking other, more sheltered areas as resting areas. The second and more likely explanation is that the insecure cat actually seeks areas and items smelling most strongly of the owners and deliberately urinates or defecates on them in an effort to associate his or her smell with the supportive smell of the absent owner. This is more an act of marking behavior and is designed to pre-sent a stronger front to any enemies, real or imagined, that may see the owner's absence as an opportunity to challenge the cat's occupancy of the home. When such elimination occurs on the owner's lap or the bed while the owner is in it, we can guess that the cat is emotionally upset for some reason and, having found protection on its mother figure, either tries to endorse the bond or simply overrelaxes and involuntarily urinates in relief.

This may be very interesting in theory, but if nothing else it demonstrates the need to isolate cause and effect. If the cat is simply trying to keep itself confident by an association of smells because it feels vulnerable in its own home when the owners are absent, then it may be better and kinder to board it in a secure cattery than to leave it at home alone. Owners sometimes have a stronger influence on the security of the cat than perhaps we realize, yet it would seem integral to the substitute mother idea. It may be possible to provide greater security by leaving the cats in a smaller area in the home, which should not include owners' bedrooms or the living room, so as to preclude the possibility of associative urination. Nervous or associative toileting should then no longer occur as the cat will be secure in the smaller, predictable area. Such techniques are also employed to help the dog that cannot cope with life alone and may suffer from similar separation anxieties. Dogs normally howl or destroy household items in response to anxiety but may also lose toilet control. It is never as directed as the cat's loss of toilet control in such circumstances and so is seen as far less of a considered action to maintain security.

If boarding such cats is not acceptable, then lodging them with friends or employing a full-time cat sitter may be the only alternative, but practicing short separations and improving overall competence at coping with novelty and challenge are vital for the permanently indoor cat. In serious cases the use of progestin treatment from the vet can help, though only for the short term to avoid the problems of water retention. Such prescription should ideally start two or three days before the intended departure so that the cat is prepared physiologically for the shock of the owner's sudden absence from home. I would again suspect that many alternative treatments could have a role to play in treating such cases but as yet have not had the opportunity to test their effectiveness.

One unfortunate aftereffect of this type of elimination

problem is that when a cat has started to soil on beds, especially on duvets, it is very difficult to break this habit. Soft materials continue to attract, and cats must be denied access to beds for several weeks to break the association— that is, if they haven't already been barred for life anyway. This should be accompanied with a litter more suited to the cat's toileting needs and the owner's expectations.

So much for the messy end of the job, though of course it does account for much of my caseload. Now we shall turn to the far less understandable or explicable aspects of feline behavior for which there is often very little on which to base treatment other than sheer guesswork. Sometimes it works, sometimes the problems go away of their own accord, and sometimes they don't. But whether treatment has been effective or not, the final chapter of this book is dedicated to those cats who demonstrate bizarre or totally pointless behavior, and also to their sometimes long-suffering owners.

᧐ 10 ᧐

Strange Diets and
Bizarre Behavior

Many of the problems that occur with cats are a result of
their excessive attachment to us and inability to function
competently without us. They also bond strongly to the
territory we provide around our shared home base. Moving
can therefore be traumatic for the cat as it involves a complete change of the whole territory, destroys walkways and
escape routes, and can leave the cat feeling totally vulnerable. Slow, careful introductions, one room at a time, and lots
of attention will help most cats over the stress of such upheaval within a couple of days, and few will need any sedative
support. It may be wise to board particularly nervous individuals in a safe and secure cattery before the packing starts
at the old house and not to bring them to the new house
until everything is unpacked and positioned. Outdoor cats
with a wider experience with change generally cope better
but should be kept in the new home for a week or so to learn
the geography and smells of their new base. When finally let
out to explore their new outdoor patch and carve out a piece
for themselves with local residents cats, they should be
hungry. Starved of food for twelve hours or so, they will not

wander too far from home and will readily respond to the call or plate bashing that signifies dinner is served. Accompanying the cat on its first few excursions into the brave new world will also help, but the cat's adaptability and survival instincts usually serve it well, and it will soon adopt a lifestyle and habits similar to those at the old house. There are exceptions, however.

Moving Home

Dear Mr. Neville,

My partner and I moved into a new apartment about six weeks ago along with our ginger cat Banzai. We kept him in for a couple of weeks, and he seemed to settle in well. However, since he has been allowed outside again he keeps returning to our old house, which is about two miles away. Is there anything we can do to tell him that we, and he, have moved?

Yours sincerely,

Madeleine Fosket

This is a very common problem. Having moved so close to the old apartment, it is highly likely that in exploring his new territory Banzai will have encountered old routes through his previous one. He will simply return home along those routes as before and then look confused on arrival to find that all has changed. The bond with the new home is simply not well enough established to attract him to it. Some are inadvertently encouraged by the new occupants of the old base, who provide food or who are flattered by this strange cat's confident entrance through the cat flap and willingness to set up home with them. But even when they have been warned that the cat might return and take repelling action by tossing him out or throwing water at him, the bond with the old center of the territory can persist. The cat will keep returning and will go to the new home only if physically taken there. Both parties can get tired of the traveling, especially in the remarkable cases where cats have returned to old haunts many miles away. Where cats have traveled thirty miles (I have encountered two of these hikers) or where epic journeys are made across the country covering hundreds of miles, there can be no possibility of this being due to encountering old routes. Cats, like homing pigeons, migratory birds, salmon, whales, and even some dogs must have a built-in homing sense that they use to navigate their way back. Most theories suggest this sense is a magnetic one and that the cat has a built-in compass, responsive to the earth's magnetic field. However, others suggest that it is more a finely tuned chemical detection system that enables a cat to detect a very low concentration of its scent or of familiar scents and follow a path toward its home, where the familiar smell or combination of smells is strongest. Treatment of these homing specialists may be impossible with our present lack of knowledge, and therefore including the cat in the sale of the first house may be the only course to take. Some have suggested attaching a magnet to the cat's head to scramble the homing sense, but I have no information to suggest that this would work. One to try for the future perhaps.

Cats such as Banzai can usually be treated successfully. The first step is to ensure that the new occupants of the old house do everything to detach him from his old home by chasing him away and throwing water at him and never stopping to say hello or feel sorry for him. Other neighbors, even if previously friendly, must be asked to behave similarly. Banzai should be kept indoors at the new house for about a month, but if he continues to return should never be taken back to the new home by a direct route. Instead the owners should make as wide a detour as possible, heading off in the opposite direction from the new home and driving a good few miles before circling around. Sometimes it even helps to board the cat for a few weeks in a cattery as far away as possible from either home in an attempt to scramble both memory of the old home and any homing mechanism. But once at the new house, again the tricks of short, frequent feeds and plenty of love and attention against a secure background should help build new bonds. The new home should come to be perceived as the center of the new territory and a source of food and shelter, in contrast to the unpleasantness of the old home. It may take weeks, and in some cases months, before the cat can be allowed outside unattended. The moral of the tale is always move at least five miles to a new home.

Where some cats bond to a territory so emphatically that they find it hard to change, others can become so attached to the owners that they find it hard to cope without them. Overdependence can lead in old age to increased attempts by the cat to maintain contact vocally. This lack of security is understandable since the cat regards its owners as protecting mother figures. The pack dog frequently suffers from separation anxieties when isolated from us, its substitute pack, and can make desperate efforts to keep with us. Trying to head us off at the door and guarding the exit and the closet where the coats are kept are all designed to maintain personal security by staying in a group. And if departure is successful, the lonely, anxious dog may howl and cry to attract us back or try to escape by digging at doors and climbing out of windows in search of us. When complete panic sets in, the dog may lose toilet control or become destructive of household property.

This is not observed in cats, yet many are clearly unhappy and insecure when separated from their owners. Some do cry a pitiful high wail or short meow, similar to the distress call of kittens. Others make concerted efforts to stay with us, the most remarkable case being that of a Burmese cat, a four-year-old spayed female and one of four in the house, that could not face life alone at night without its owner. This cat scaled the outside of the house via the gutters and ledges to the eaves of the roof outside the owner's bedroom, three stories up. It crawled in through a hole and then wailed to attract attention. This happened every night for about three weeks. The owner naturally responded to the awful din by getting up in her nightdress to rescue her via the attic and then keep her with her in a nice, warm bed all night. This is a typical case of learned incompetence, now thankfully resolved by a little social detachment between owner and cat, a restructuring of their relationship so that affection is initiated only by the owner and never the cat, and mainly by offering the cat an extremely attractive heated bed downstairs in the kitchen. The case served to indicate how strong the bond can be between solitary predator and owner, when we provide warmth and protection.

Dear Mr. Neville,

Occasionally my wife and I have to go abroad and, because of the quarantine regulations, are forced to leave our beloved cat, Shona, behind. Usually we take her to a cattery of impeccable hygiene and comfort but find that she is most distressed when left. She will barely eat, spends much time in her bed, and, if pressed into a little activity by the cattery staff, may even lash out at them. Yet as soon as we arrive back to collect her, she comes running like a dog to see us and even seems to recognize the note of our car engine. She purrs furiously and immediately settles back into our arms. Once home again, she eats furiously and follows us around the house as she normally does, apparently none the worse for "vacation." Can we do any more to help her bear the separation better?

Yours sincerely,

George and Julietta Barclay

Whether cats like Shona are highly attached to their owners or simply dependent on the whole nature of their lifestyle is sometimes difficult to assess. Most owners would rather believe that it is they for whom their cat pines when they leave it in a cattery. This feeling that the cat is distressed by their departure compounds the burden of guilt that owners bear as a result of being unable to take the cat with them on vacation. No matter how much time and effort cattery owners invest in making their establishments clean and attractive, owners may always tend to regard them as temporary prisons. The strength of the bond between the cat and the owners can be at the basis of the cat's security to such an extent that separation anxieties result when the cat is taken away to even the most sumptuous of catteries. Some boarding kennels are now trying to offer the creature comforts of home for boarding dogs with chairs and beds to climb on, televisions to watch, and high staff-to-dog ratios to keep them amused. Perhaps this will start to happen soon with cats, though I doubt it will bring much improvement for cats like Shona, whose home, routine, and owners are such a vital feature of their well-being that all else is inadequate and frightening. Treatment of such occasional anxieties is very difficult, but some progress can be made by preventing the cat from following its owners so much when it is at home, and being generally less responsive to the cat's demands for attention. More important still is for the cat to go to the cattery for a few hours on a few occasions before the real boarding period so as to get to know the cattery staff while the owners are present. All comforts from home such as beds, favorite toys, and old clothes smelling of the owner can also help the cat transfer its loyalties, though this is likely to have limited success and will probably be possible with only one particular member of the staff. The most acceptable person should also be encouraged to visit the cat in its own home to make friends and take over the role of feeding and fussing. Meanwhile the owners become more aloof. The aim is for the cat to develop the ability to cope with the change of home and lifestyle and to be assisted by the new temporary caretaker. It may be possible in only a very few catteries with very special staff, and far more cats simply just have to suffer from their separation anxiety. Whenever possible, such cats will usually be happier if left at home to be cared for by a cat/house sitter or "pet aunt" who will call to feed and care for the cat. At least the cat's whole world is not disrupted, and it can continue to be surrounded by familiar territory and comforts, if not its owners or familiar routine.

On returning to a cat left behind, the owners will often be greeted by a huge display of feline greeting with little yells, purring, and rubbing as the cat rediscovers its source of maternal security. We respond to a cat's display of affection or flirtatious greeting postures by issuing a stream of high notes in the same way mother cats and their kittens communicate with each other. This can occasionally have amusing repercussions.

Musical Cats

Dear Mr. Neville,

My cat Mignon loves everyone. She rolls on her back invitingly and adores being stroked and fussed over by any visitor or even people she meets in the street. She is always purring and is quite the happiest cat I have ever known. If I whistle, she comes like a dog and starts to roll, purring loudly, but if I whistle high notes she leaps to her feet and runs up my body to rub around my mouth. She grips hard with her paws around my neck and rubs continuously with the sides of her mouth against my face and under my chin. Sometimes it's easier not to whistle at all, but as a musician I simply must practice my flute playing. This sends Mignon into a rapturous frenzy as she leaps onto me, digging her claws into my leg and rubbing furiously around my face and the flute. It's quite impossible to practice, but if I shut her out she claws at the carpet by the door, desperate to get in with me again. Is she just a musical cat, or are there logical explanations for this strange behavior?

Yours sincerely,
Jennifer Hillborough

The high notes of our whistle or, even better, the purer, higher ones of the flute or other lighter wind instruments, seem to mimic the sounds made from a kitten to its mother. Having attracted the mother, the calls cause her to rub the kittens and physically reassure them by licking. This often happens around the face of the kitten. Mignon is probably seeking face-to-face contact with her flute-playing owner because the sounds, no matter how well practiced, indicate distress, and she wants to offer her owner reassurance of the maternal kind. This role reversal in our usual relations is unlikely in most cats. One would suspect that queens that have had litters of their own or particularly friendly cats used to much vocal communication would be most likely to respond in this way.

We all talk to our cats in high-pitched tones because they seem more responsive to them than to our normal voice or low-pitched noises. Quite apart from the felinese gobbledygook that we spout to attract their attention, we also use very high, squeaky "come hither" sounds, and the cats respond once again as if they were kittens responding to their mother's call. They trot toward us with tails high, expecting to be greeted and petted by a friend, and we keep up this form of greeting throughout their lives. The combination of sounds is almost impossible for us to detect, but as cats will usually not respond as readily to strangers making similar noises and offering similar affection we can presume that each of us has our own personal squeak, recognized by our own cat and few others.

The hearing ability of the cat is approximately the same as ours at low frequency but far superior at high frequencies. We can hear little above 20 kHz, and a dog, despite a reputation for good hearing, detects little above 30 kHz. But the cat responds to notes of 65 kHz, at least one-and-a-half octaves beyond our limit. So our high squeaks are well within their range, and their general ability to detect such high notes may account for the jumpiness of some cats around our household electrical appliances, such as the washing machine.

However, for super-responders like Mignon there may be no alternative but to banish her outdoors or two rooms away during flute practice, or else there should be a little less tolerance of her efforts to reassure the owner when she is emitting those distress calls.

Again we tend to suffer more from dogs that demand attention than cats. The demanding dog can employ a variety of ruses to gain center stage, from standing in front of us wagging its tail to leaping up at us, barking, and carrying off slippers, wallets, and handbags as trophies. Cats are usually more subtle and calmer with their demands, though a few will run up legs to get to the cuddling zone. Fine if you're wearing jeans; painful if you're in a skirt and tights. Most cats will seize the opportunity for a little maternal comfort and affection by jumping on our laps whenever we sit down, recognizing that we are then available, relaxed, and most likely to be responsive to demands for attention. Failure to respond on our part usually causes the cat to move away in a state of obvious huff, its back turned and head held high and aloof. A few go a step further and insist that we provide some form of attention by protest-spraying (see page 146), and a few others will engage in other activities designed to make us scurry toward them. A favorite trick by these cats is to leap onto windowsills and tread less than gently through our favorite potted plants or most prized ornaments and antiques. It works every time. Thankfully most demanding cats stick to repeated efforts to climb on our laps or simply paw at us and, when met with no response, simply pretend that they don't care anyway.

The nature of our relationship with the cat has been referred to as maternal throughout this book, and never is this more apparent than with the many cats that, even in the prime of adult life, will readily suckle on our clothes and occasionally even our skin. They salivate profusely during such behavior in anticipation of the milk feed that it used to bring them as kittens, while their paws and claws make a stepping, contact-and-relax motion that used to be directed at the mother's nipples to stimulate milk flow. With a big, bruising adult cat this behavior can seem incongruous with its outdoor despotism toward all other life forms and not a little painful and damp for the owner/victim. This behavior is seen in nearly all pet cats to some degree and is triggered by our physical stroking and handling or simply by our reassuring presence. While this may be tolerable, the cat that dribbles all over its owner is usually allowed near for only short periods and tossed off when the saliva starts to flow. Kittens that have been hand-raised may demonstrate this behavior more readily and with greater gusto for the obvious reason of full maternal substitution.

Some owners, particularly with rather nervous cats or those rescued from an unfortunate beginning, may be unwilling to reject the cat for fear of contributing to further rejection or withdrawal of care. Undiscouraged, the cat may continue with this behavior to the point of bringing its owners out in rashes or rendering corners of their clothes wringing wet and well chewed. While it is easy to suggest that such cats be discouraged from perpetuating infantile behavior patterns, there may be a strong reluctance on the part of the owner to reject the cat by putting it down onto the floor or by being less available for any affection. Keeping encounters short and frequent, with adult interactions of rubbing and stroking, soon replaces the unwanted infant patterns if the owners so desire. Others, particularly those more dependent than most on their cat for company, such as many elderly or housebound people, actually prefer the behavior to persist and are prepared to bear the damp consequences at the price of having something warm and responsive to care for.

Such cats seem especially interested in sucking woolen items of clothing. This may be because the texture of wet wool may be similar to that of a mother cat's fur around the nipples or more likely because the smell of lanolin and other natural odors in wool are similar to those in the nursing cat's fur. But wool sucking is a different behavior from the wool eating that I have encountered in many oriental cats. Wool and fabric eating or the eating of any nonnutritional item is termed *pica*.

Pica—Fabric and Wool Eating

Why some cats should want or need to eat wool, and indeed other fabrics, is not understood. That they do is beyond any doubt, as many owners of clothes, carpets, and furniture covers with holes in them can testify. In the 1950s the behavior used to be thought of as a trait restricted to certain strains of the Siamese breed and particularly of queens in estrus. As it seemed to be an inherited disposition, some attempts were made to breed it out of these strains. The results of a survey that I have been carrying out since 1988 reveal that the problem is more widespread and not restricted to the Siamese. Of 152 cases reported at the time of writing, 59 percent are of the various types of Siamese, 28 percent are Burmese, with 13 percent of mixed parentage. Some of this latter group have one parent Siamese or Burmese, but some do not and are apparently pure moggy. While their exact family history is unknown and many contain some orientals, it is likely that the condition simply occurs less frequently in crossbred cats. It is perhaps more logical to think in terms of some genetic abnormality caused by malfunction at the chromosome level, which is more likely to occur in cat types that have been closely bred. Of those responding to the survey who had contact with owners of their cat's littermates, 58 percent said that at least one of their cat's brothers and sisters was also a fabric eater. Not conclusive, but it does suggest that the condition is inherited at least to some extent.

The most likely explanation for fabric eating probably lies in a malfunction between the brain and neural control of the digestive system, though the degree of involvement of the central nervous system (voluntary decision making) and the autonomic nervous system (involuntary nervous action such as control of digestion) is as yet unclear. Also the relative frequency of this behavior as shown in the survey results is probably more a reflection of the willingness of certain owners to reply to the questionnaire than a measure of true occurrence.

Dear Mr. Neville,

My Siamese cat eats not only wool, which we keep well out of her way, but also all other fabric. She has a particular passion for towels and tea towels, socks, and T-shirts and has now started chewing at the settee cover. She will chew in front of me, and even on me if she gets hold of my sweater, and seems unperturbed by our screams of disapproval and even squirting water at her. If we go out, I now lock her in the kitchen, where she can't get at anything. Is there anything we can do to stop this expensive behavior?

Yours sincerely,

Terrie Peters

Some wool/fabric eaters do stick to consuming wool, and again one might expect that the smell or texture of wool is the initial trigger to the behavior. But the majority broaden their appetites and will consume all fabric. Items of clothing, preferably worn, and towels are especially popular. In the survey 93 percent consume wool, 65 percent will eat cotton, and 54 percent synthetic fiber. Perhaps the scent of our clothes attracts the cat and stimulates the need to consume them. Why two cats in the country at least should restrict their fiber eating to the underpants of the lady of the house is unclear and too rude to investigate further. In one of those cases the last ritual at night before retiring is to locate and remove the cat from the bedroom, where it has cunningly concealed itself in the hope of a successful abduction and consumption of the discarded panties.

When eating any fabric, be it underwear or less personal items, the cat will appear totally engrossed in its activities and will sometimes be in a trancelike state. Intervention by hissing or yelling or even throwing water at the cat may cause it to stop, but often it will simply go straight back to the item or look for another in a quieter place. The cat will take in the wool with its canine and incisor teeth and, having obtained a good mouthful, will start to grind it up using its shearing molars at the back of the mouth. The volume of fabric consumed by some cats is truly remarkable, and even more so when one considers that it usually passes through the cat unaltered without causing any harm. Unfortunately some do suffer blockages in the stomach or farther down the digestive system, and a few are euthanatized because of the resulting damage, but some live long, healthy lives, eating wool or other fabric every day without any repercussions.

A few more are destroyed because their owners find their habits too expensive to live with. Some have damaged dresses worth hundreds of pounds or furnishings worth thousands, and the average is £136 (approximately $375) per fabric eater. Interestingly most who replied said they had learned to live with the problem and would not dream of parting with the offender, nor would they be deterred from having a cat of the same breed in the future.

One theory as to the origins of fabric eating suggests that, like wool sucking, the behavior is linked to a continuing infantile disposition in what are traditionally sensitive breeds. Thirty-nine percent of cats in the survey exhibited the usual forms of continuing infantile patterns such as excessive kneading of their owners, suckling and salivation when petted, and perhaps an overattachment expressed by the need to stay in physical contact with them. The age of weaning of kittens may also be crucial. Kittens taken from their mothers at six to eight weeks of age may be more likely to show stressful and nervous responses later in life than those taken at twelve weeks. The extra weeks, though perhaps unnecessary from the nutritional point of view, may be essential for the emotional development so that it can later cope with challenge more effectively and so be less likely to reach an emotional threshold for fabric eating.

Yet 61 percent of cats in the survey didn't exhibit other continuing infantile behavior patterns, suggesting perhaps that while we may encourage the development of wool eating, the underlying disposition is unaffected by our relationship with the cat.

Seventy-eight percent of cats surveyed started to eat fabric during adolescence at four to twelve months. Twenty-two percent began in response to some form of stress or trauma, such as medical illness, moving, or the acquisition of another cat, so perhaps the notion of the phenomenon being an inherited disposition requiring an environmental or emotional trigger is deserving of further investigation. It is apparent that, once started on wool, most move on to other fabrics. Males seem just as likely to start as females, but neutering alleviated the behavior in only about 15 percent of cases. Some cats cease eating fabric altogether at about two years of age, perhaps because they learn to cope with the challenge or change that triggered the behavior in the first place. A few are treatable, for instance those that have been only slightly pushed over the reaction threshold, though quite why some should respond and not others is a mystery.

My first approach is to consider the prospect of overattachment being an important influence when its withdrawal or lack of constant availability of the owners leads to a form of stress-avoiding fabric eating. Such cases usually concern cats that are only found to eat wool or fabric in their owners' absence but are often clingy and dependent when they are around. Treatment for them is encouragement to grow up and to exchange any continuing infantile reactions with the owners for more adult ones and to keep affection in short doses. Wherever possible these cats are encouraged to go outside to further their level of stimulation, reduce the importance of owners and home for activity, and help establish a less dependent relationship with the owners. Many of these problems appear to resolve, though how many are finding fabric sources in the homes of neighbors or from their clotheslines is difficult to say.

Access to edible fabric must be made impossible for all fabric eaters, and sometimes denial for a few weeks causes the behavior to cease. Direct negative conditioning by ambushing the cat with a water pistol while in mid-chew has helped some, though often such tactics only produce a secret fabric eater. Indirect tactics using taste deterrents applied to specially laid towel baits can have a dramatic effect in deterring the cat from eating all fabric forevermore, though choice of deterrent is crucial. The traditional tastes of pepper, mustard, and chili or curry are invariably useless and simply broaden the cat's desire for a more exotic diet. Aromatic compounds such as menthol and oil of eucalyptus seem to be more successful and have reformed a number of ardent fabric eaters.

The best hope seems to lie with dietary management, though why this should be is difficult to explain when most fabric eaters have a perfectly normal, healthy intake and their appetite for it is undiminished by having a stomach full of acrylic sweater or woolen scarf. Providing a constantly available source of dry cat food, in addition to offering the usual diet, can redirect the desire to eat fabric to more nutritional targets, apparently without risk of weight gain. Most cats simply snack all day on the dry food and cut down voluntarily on the intake at usual mealtimes. Sometimes it helps to cease mealtimes altogether and simply leave a constant supply of dry food for the cat.

If keeping the stomach constantly active and partially full switches off the need to eat fabric in some sufferers, it may be due to the pleasurable, comforting feeling of having food in the stomach rather than as an alternative form of intake. This may explain why adding fiber, which helps pad out the volume and passage time of traditional canned diets, can help prevent fabric eating. Padding such as bran can be added to the usual diet up to a point, but most will not accept too much. Instead adding small lengths of finely chopped undyed wool or tissue to the diet may be more acceptable to the cat. This is a form of giving in, though it can be a lot less expensive than letting the cat select its own fabric from the wardrobe. Other owners have resolved the problem of indiscriminate fabric eating by providing the cat with a towel to chew at dinnertime. The cat takes a few mouthfuls of food, then chews and eats a portion of towel—a truly bizarre spectacle, but extremely effective in some cases.

With some fabric eaters the time taken to ingest food

seems to be relevant to the frequency and severity of their fabric eating at other times. In the wild, cats would have to stalk, capture, handle, and kill their prey before consuming it. Consumption itself would take some time as the cat would have to rip through the fur or feathers to expose the flesh. Such gory necessities are not precursors to eating for the pet cat. We hunt its food for it at the supermarket or pet shop, wrestle it from the cabinet, render it dead by piercing the box or cutting its lid off with a can opener, and expose all the good portions on a plate ready for the cat to consume at its leisure. No wonder so many cats have bad teeth—they never get used. If through the evolutionary development the cat has come to be so programmed as a hunter as to require the capture and prey element as a precursor to stimulating appetite or correct digestion, then the cat will benefit by being forced to invest more time in processing its food. This certainly seems to be the case with some fabric eaters that are forced to gain their food by spending much time chewing at the gristly meat and indigestible sinew attached to large bones rather than being offered free meals on a plate. Providing tough chunks of meat forces the cat to invest more time in its total feeding behavior and helps reduce or alleviates entirely the desire to eat fabric at other times.

More than this I do not have to offer the owners of wool-eating cats at present, though the results of the survey and details of the cats' pedigrees, where appropriate, are currently being processed further by fellow behaviorist Dr. John Bradshaw and his student Diana Sawyer at the University of Southampton. Hopefully this will shed a little more light on the origins of this strange phenomenon. We hope then to carry out more direct observational and analytical trials, especially on the cat's dietary and breeding aspects and on the influences of stress on fabric eating, in collaboration with Dr. Tim Gruffydd-Jones at the Department of Veterinary Medicine at Bristol, where I hold my monthly referral clinics. Tim is one of the leading authorities on feline medicine in the United Kingdom, and so with all angles covered we hope to enlarge our understanding of fabric eating in cats quite markedly over the coming year.

Pica—Electric Cords

Dear Mr. Neville,

My Burmese cat, Rangoon, has a very dangerous behavior problem. He loves to chew electric cords, but thankfully so far has yet to connect to the wires inside. I've tried telling him off, but he seems incurably fascinated by them. Is there anything to do, or is he doomed to be electrocuted one day when I'm caught napping?

Yours sincerely,

Siobhan O'Flaherty

This apparently self-limiting behavior could obviously be fatal to the cat and also to the owner and her property because of the risk of fire. Understanding not one jot of the motivation to chew electric cords, I took the view that an early interest in a movable play cord had developed into predatory handling, the end sequence of which is to consume the dead prey. So while some cats simply hunt mice, process them ready for eating, and then lose interest prior to ingestion, Rangoon jumped from stalking to consumption and learned to miss out the handling sequences. If it was a learned behavior pattern, perhaps a little negative conditioning could undo it, along with the offering of suitable alternative interests. Just for good measure we began by treating Rangoon as a fabric eater and altered his diet and feeding patterns in the hope of switching off any predatory or appetite-motivated aspects of his behavior. We tried to redirect his play and mock predatory activity by encouraging him to spend more time outdoors and with an attractive range of toys, some laced with catnip, which fortunately proved irresistible to Rangoon. Then we allowed unsupervised access to electric cords suitably coated with eucalyptus oil and unplugged from the outlets. Something from this battery of suggestions was successful as Rangoon no longer chews electric cords and leads a happy, safe life. Friend and cartoonist Russell Jones decided that Rangoon was actually trying to administer electro-convulsive therapy to himself for some other unseen psychological disorder, and in the absence of any other serious explanation for this behavior (which I have encountered in only two cats, both Burmese) I almost have to agree with him.

Pica—Rubber

A good number of the owners who responded to my survey on fabric eating also mentioned that their cat ate rubber. Most restricted themselves to the delights of rubber bands with not a little risk of gastric problems. However, Wilberforce specialized in condoms. His career of lurking in the bedroom came to an abrupt end when, the night after his owner had overcome his embarrassment and called me, Wilberforce made the mistake of trying to capture his prize after it had been fitted but not yet employed for its intended function. This was the last straw for Wilberforce's owner, who, doubtless smarting in pain, took Wilberforce to the animal shelter the next day. Quite why some cats, particularly oriental breeds again, should be so obsessed by rubber I do not know, though they are often attracted by plastic as well. Treat as for electric cord eaters and pray!

Plant Eating

Most cats eat more plant material than we realize, probably in an effort to obtain a quickly digestible source of vitamins and minerals and roughage. Some regurgitate the grass with a portion of their dinner, and this is believed to be a natural method of worming or helping to eject hair balls. Leopards can live for many weeks on plant material alone without ill effect, though all cats are obligate carnivores and cannot be maintained on a vegetarian diet for very long. Indoor cats should be provided with a tub of seedling sprouts to munch on so as to discourage consumption of houseplants. These are available from pet shops and, being more attractive than most houseplants, are readily consumed in preference. A little negative conditioning with a water pistol can help convince the cat obsessed with eating one's favorite plant, but this is difficult when the plant in question lives in someone else's garden.

One cat, Suki, caused great concern because she had to undergo surgery six times to remove a jagged edge of a leaf of one particular plant from her throat. Identification and preferably removal of the plant is the obvious course of action to take, but this has so far proved impossible. The cat seems to have an irresistible urge to eat only that plant, though we are presently hoping that a few weeks' confinement indoors after the last operation and the opportunity to eat seedling sprouts and a dry diet may help redirect the cat's desires and alter her habits. If we could only take an identifiable sample of the plant, we may find that, like catnip, it contains some irresistible compound that could then be offered to the cat in a safer form.

Other interesting cases have concerned cats with passions for eating certain foods or simply all food. Polyphagia, as this is known, is not at all common in cats, as they are renowned for limiting their intake of food to suit their needs. One does encounter the occasional feline barrel, however, and aside from the many medical conditions that can cause an increase in appetite, this is most frequently observed in permanently indoor cats. Boredom eating enables the cat to occupy itself, and as with humans, eating seems to help relieve stress. Treatment usually involves providing more carefully controlled portion feeding and denying access to food at other times, coupled with greater stimulation and activation of the unoccupied cat. Outside walks and acquiring a second cat can help dramatically in the treatment of boredom eating.

An Alcoholic

Quite what to do about a client's cat with a passion for Campari and soda was a little more difficult. Gradual withdrawal of access to alcohol was advised as a slow drying-out program for Slosher, who had been used to a daily drink for over three years. I also advised that the soda level be raised at the expense of the Campari and hoped that cold turkey would not set in.

Sexual Problems

And so finally to sex. Sadly this is a short and disappointing section as cats hardly ever present sexual problems and are no match whatsoever for the often randy and excessive exploits of the young male dog. Females are far less likely to show unwanted or bizarre sexual behavior in cats or dogs, and the fate of most male cats is to be castrated at an early age. On entering adolescence and perhaps just prior to castration (or even the cause of it) toms may occasionally mount their owner's arm, cushions, or furniture and masturbate, particularly following a period of stroking or excitement, but the behavior is usually short-lived and bears none of the insistence of the randy young terrier's assault on one's leg.

Nor are cats famous as flashers like so many dogs, including my own Colonel, the Doberman, whose earlier name prior to the operation was Pencil. Indeed a cat's sexual responses seem unexpressed except under the single stimulus of the smell of a queen in season. The unneutered male can detect this from miles away and will pursue with the utmost vigor. But courtship is usually short and would be looked upon as rape were it not for the willingness of the queen to adopt the swaybacked receptive mating position. The performance is usually very quick, though it may be repeated a number of times despite the fact that withdrawal is a painful affair for the queen, due to the backward-facing barbs on the tom's penis. This stimulation of the walls of the female's vagina, while painful, stimulates ovulation, thus ensuring that every mating has the best possible chance of success. It also means that kittens in a litter may have different fathers if several suitors are at the scene when the queen is receptive.

A cat's sex life is predictable and neat, and I have dealt with only two cases of misdirected sexual behavior. One concerned a totally inexplicable case of a three-year-old neutered male cat who persistently mounted one particular radiator knob. It remains high on my list of favorite self-resolving problems, though I am none the wiser as to its origins. High on my list of treated problems is also:

Dear Mr. Neville,

Over the last few days Thumper, my neutered ginger tom, has fallen in love with Jasmine, my eight-year-old dachshund. He follows her everywhere, yowling and sniffing, and tries to mount her at every opportunity. She is clearly distressed and tries to hide, but to no avail. He is like a cat possessed and has kept up his advances almost continually. What on earth is happening?

Yours sincerely,

Richard Calvin

As it turned out, poor Thumper was responding quite normally to a series of unusual pheromones being given off by Jasmine as a result of an internal tumor. These triggered his sexual response in a similar manner as do the smells of a queen calling, even though he was neutered. The reception areas of the brain remain unaltered by castration, and ordinarily the normal responses are prevented in the absence of male hormones. Occasionally other hormones can compensate for the lack of androgens and engender sexual responses. While it is not uncommon to see castrated dogs continuing to show sexual responses after castration, it is very rare in cats. Thumper was probably one such exception and could do nothing else but respond to the plainly enticing smells of Jasmine. The vet treated the dog, and the cat's behavior problems disappeared overnight!

Epilogue

If this journey along the roads of feline insanity and undesirable behavior has helped even one owner enjoy and understand his or her relationship with a pet cat a little more, then its aim will have been achieved. My own cats, Scribble and Bullet, have joined me from time to time during the writing of this book to gaze down at me from the top of my word processor and form their own opinions about why their previously doting mother-figure owner should abandon them for the sake of tapping away on a keyboard for hours on end. Perhaps they will produce the sequel to this book and call it *Do Owners Need Shrinks?* The answer to that, and to the title of this book—*Do Cats Need Shrinks?*—is YES! Not all the time perhaps, but occasionally and definitely YES!

If your cat is displaying a behavior problem, first seek the advice of your vet. Your vet may then refer you to another vet who is a practicing animal behaviorist. Help *is* available!

My practice is one of the original members of the Association of Pet Behaviour Consultants, which may be contacted by writing to the Honorary Secretary, The Association of Pet Behaviour Consultants, 50 Pall Mall, London

SW1, England. The American Veterinary Association of Animal Behavior is a similar organization in the United States, and it may be contacted by writing to Dr. Wayne Hunthausen, Secretary/Treasurer, The American Veterinary Association of Animal Behavior, 4820 Rainbow Boulevard, Westwood, KS 66205.

Glossary

airing cupboard: A closet built around a hot water pipe or tank so that newly washed clothes can be warmed there.

bolt hole: A safe hiding place.

cat flap: A cat door; a small door that allows a pet access to and from home.

Elizabethan collar: A funnel-shaped "collar" that fits around an animal's neck and head to prevent it from pulling out stitches or grooming an infected area.

garden: A yard with plants.

kittening pen: A box or enclosure for a mother cat and her kittens, usually with edges high enough to keep the kittens in but let the mother cat out.

litter tray: A litter box.

loo: A bathroom.

moggy: A housecat.

patch: A cat's area or territory.

veterinarian surgeon: A veterinarian.

yard: A small paved area adjoining a house.

Fat & Bones and other stories

Fat & Bones

and other stories

LARISSA THEULE

illustrations by
ADAM S. DOYLE

CAROLRHODA BOOKS
MINNEAPOLIS

To Philip, whose love and support
never falter —L.T.

To those light of heart whose
blood still pumps a good dose of
mischief —A.D.

Text copyright © 2014 by Larissa Theule
Illustrations copyright © 2014 by Adam S. Doyle

Carolrhoda Books
A division of Lerner Publishing Group, Inc.
241 First Avenue North
Minneapolis, MN 55401 USA

For reading levels and more information, look up this title at
www.lernerbooks.com.

Main body text set in Janson Text LT Std 11/17.
Typeface provided by Adobe Systems.

Library of Congress Cataloging-in-Publication Data

Theule, Larissa.
        Fat & Bones / by Larissa Theule ; illustrated by Adam S. Doyle.
            p.   cm
        Summary: When a farmer dies, the decades-old feud between his
    son, Bones, and Fat, a fairy who lives in a tree on the farm, escalates, with
    surprising consequences for Bones's mother and the farm animals.
        ISBN 978–1–4677–0825–8 (lib. bdg. : alk. paper)
        ISBN 978–1–4677–4623–6 (eBook)
        [1. Vendetta—Fiction. 2. Domestic animals—Fiction. 3. Fairies—
    Fiction. 4. Farm life—Fiction.] I. Doyle, Adam, 1975– illustrator.
    II. Title.
    PZ7.T3526Fat 2014
    [Fic]—dc23   2013030364

Manufactured in the United States of America
1 – BP – 7/15/14

CONTENTS

Fat and Bones

Fat stood on the topmost branch of
the tree, gazing in the direction of the
farmhouse. Something had happened. He
could feel it in the tips of his blue wings.
The farmer's wife had not yet made a trip
to the pigpen, and already the afternoon
sun had begun to lick the fields. A slight
shiver passed through Fat. There would be
a full moon tonight.

Fat sniffed the air and smelled something
putrid. Something else too. Something faint
and pleasant, like new seedlings coming up
from the dark earth.

Change. That's what the fairy smelled.

Fat adjusted his belt. He had observed lately that Bald had grown stooped and weak. The old farmer would drop tools and 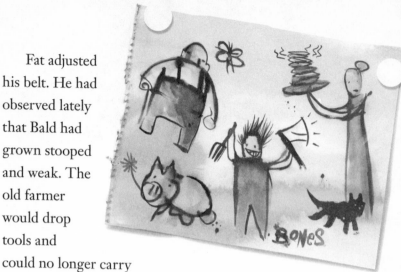could no longer carry heavy machinery. One time, a pig had broken loose from the pen and pushed Bald over so the old man's crooked legs waved in the air.

Lately, Bald's son Bones had taken over the farm. Fat hated Bones, as Bones hated Fat. Although they'd never spoken to each other, their hatred had grown dense and deep, too thick and round not to roll over everything in its path. Nobody knew the reason why. Some said it was because Bones had shot BBs at Fat when Bones was younger, and Fat had responded by dumping a potion in Bones's coffee that turned his hair into bramble. Or they might have hated each other because one wished for wings and the other wished to be tall. Whatever the reason, nothing would do but for their hatred to burn itself out.

Fat sniffed again. Death. That was the change in the air.

The door to the farmhouse opened. Bones carried out the body of his dead daddy and dumped it into a crude hole. He began to shovel dirt on top. When Bald was covered up, Bones dropped the shovel and turned to Fat's tree, the only tree in the field. Bones grinned. It was the kind of grin that could make one's stomach shrivel like a raisin. Bones had buried peace along with his daddy. Bones wanted war.

Fat smiled. War would be a nice change of pace. To create a bit of disorder, a little chaos. Quite a welcome change, really. He went into his hole to plan.

✳ ✳ ✳

Back inside the farmhouse, Bones patted his weeping mother on the back. He himself felt no sorrow for the loss of his father; his heart was too small. "Don't worry yourself, Ma. I'll take care of you."

Bones had no idea how to take care of himself, let alone his mother. In fact, he only noticed her when she placed a plate of food in front of him. Even then, his acknowledgment of her came in the form of a hungry grunt.

Mrs. Bald rocked harder in her chair, enormous tears tumbling down her cheeks.

"Ma, please, get over it," said Bones in as kind a voice as he could manage. "Pa's dead. Dead is dead. No point crying about it."

Mrs. Bald wept louder.

"C'mon, Ma."

Mrs. Bald's tears began to soak her hair and clothes.

"Ma, put a sock in it!" roared Bones. "I'm hungry. Fix me a pot of pig foot stew. That'll set you right."

He grabbed his hat and charged through the front door, kicking aside the cat's shiny red food dish on his way out. He had planned on waiting until the following day to chop down Fat's tree, but Ma's crying grated on his nerves. A good swing of the axe was just what Bones needed. A good swing of the axe would get rid of that wretched fairy, and all would be well.

Hate has a habit of rendering its keepers blind. Which is why Fat did not look down to see Bones lumbering through the wheat toward his tree, never mind the axe in Bones's hand. Nor did Bones look up to see Fat navigating the air currents above him.

Bones reached the tree. Without stopping to plant his feet, he swung the axe, lost his balance, and fell to his knees. A strand of wheat whipped across his face. He growled and picked himself up. The next time, he stood firmly, his feet a stable distance apart. He swung the axe across his shoulder, gritted his teeth, and *whack!*

A shriek of pain pierced his ears. The farmhouse cat had streaked in front of Bones precisely at the moment he swung the axe, and Bones had chopped off its tail. The cat howled and ran around and around the trunk. Four inches of severed grey tail hung from the tree, wedged into the cut made by the axe.

First, his mother, now, the cat. "Stop crying!" yelled Bones. "It's only a tail!"

Bones decided to shut the cat up for good. Each time the cat circled the trunk, Bones swung his axe again, but each time, Bones missed. After a while, the trunk of the tree was riddled with axe marks. Bones was as far off from felling the tree as when he had arrived, though now his arms were nearly limp.

When the cat grew dizzy and wobbled off into the wheat, Bones leaned on the axe handle and wiped his brow. He no longer had strength left to chop down the tree. He kicked at its trunk instead and broke three toes.

Meanwhile, Fat flew through the open door of the farmhouse. Bones had forgotten to close the door on his way out, and Mrs. Bald, still weeping, was in no frame of mind to have closed the door herself.

Fat hovered above the pot of pig foot stew on the stove. He held a bottle of Sunflower Skeleton Eraser in his hand. He would drop it in the soup. Then, when Bones ate his dinner, what a sight! Bones would turn into a blob of flesh on the floor—a rubbery rug—a boneless Bones!

Fat swooped and dipped in front of Mrs. Bald's face as she stood stirring the thick brown stew. She did not even blink. The woman couldn't see him through her tears. Fat giggled and said, "Get ready, bonesy Bones."

He pulled and prodded at the stopper wedged in the bottle, but the stopper would not budge. He bit his lip, and sweat formed on the brow of his round head. He was concentrating so hard that he flew into the wall. He dipped and fluttered, holding his head with one hand, the bottle with the other. The bottle slipped, dropping into the boiling stew with a delicate *plop*.

"Aaaaah!" Fat squealed.

Mrs. Bald raised her head. "Bones, is that you? I've made pig foot stew, my boy, just like you wanted. It was your father's favorite too." At that, she began to weep all the more.

Fat did the first thing that came to mind. He dove headfirst into the stew. The mealy liquid grated against his skin. He struggled to remain clearheaded. He forced himself to open his eyes and scarcely avoided ramming into the pig foot itself. The foot brushed against his arm, squishy from the heat, and then floated away.

Mrs. Bald's wooden spoon came from behind and began stirring Fat around the pot. Nearly unconscious from holding his breath, Fat began to prepare for an undignified death as bits of vegetables and stew grit hit his face.

Then, through the murk, he spotted an elegantly scrawled *S* and lunged for it. His hand closed around the bottle. With the last of his energy, Fat climbed up the wooden spoon, gasping for air once he broke the surface. His face red and his blood vessels nearly popping, he burst from the pot with a mighty *schloop*.

"Bones? What are you up to? Wash up now, the stew's nearly ready," said Mrs. Bald, blowing her nose into her red-and-white checked apron.

Fat freed the bottle stopper with his teeth and spat it out over his shoulder. He turned the bottle upside down and dumped the Sunflower Skeleton Eraser into the pot.

Mrs. Bald smoothed her apron and picked up her wooden spoon. She stirred the stew once more and then lifted the spoon to her mouth for a taste test, licking it all over with her long pink tongue.

"No!" cried Fat. He had not intended Sunflower Skeleton Eraser for Mrs. Bald. If Bones found his mother flattened on the floor by the stove, he might figure the stew was to blame and not take a bite.

But it was too late.

The cat streaked in from outside, trailing blood from what was left of its tail. Moments later, Bones entered, his axe slung over his shoulder.

In less time than it took for Bones to hang up his hat, Mrs. Bald had dissolved into a mass of skin, fingernails, and hair, lying helpless on the floor. Only her eyeballs remained their original shape.

"Ma?" said Bones, poking her with his finger. "How'd you get so flat?"

Mrs. Bald looked from him to the pot on the stove.

"Did you burn yourself flat?" asked Bones.

Mrs. Bald's eyeballs swiveled in their sockets.

"Did you cry yourself flat?"

Mrs. Bald wheezed.

Fat chose this moment to take flight. But the stew had begun to congeal on his wings, and he moved slowly.

Bones saw Fat floundering through the air. He looked

at the pot on the stove. Though not usually a very clever man, he put two and two together.

"You dirty rotten fairy!" He lunged. He reached. He grabbed. He missed.

Although Fat would have preferred to squeeze out a triumphant cackle, all he could manage was a toadlike croak. The hardened stew had thrown off his navigation skills. After banging into the doorpost, then banging into the post again, he escaped into the night air.

"I'll skin you alive, you fat devil!" screamed Bones.

The tailless cat sat on Mrs. Bald's throat, licking her face.

※ ※ ※

Mrs. Bald, lying helpless on the kitchen floor, was in danger of drowning in her own tears. She had cried the night through. Bones discovered this when he awoke the next morning and stepped from the bottom stair into an inch of water. He was not clever enough to think what to do with her. Even now, standing over her with his hands on his hips, he could think of only himself and his empty stomach.

"Where're my pancakes?" he said.

Only then did Bones realize that if he did not stretch his mother out or hang her up somewhere, she would drown, and he would never eat again. Under the threat of starvation, Bones hoisted his mother over his shoulders and transferred her to the clotheslines outside. He secured her on the line with wooden pins.

Bones's stomach growled. He wanted pancakes. He wanted pancakes so badly that he went to the garage and got the tire pump. He stuck the nozzle into his mother's mouth and held her lips close around it. Then he pumped. He pumped and pumped and pumped, but she did not inflate. Instead, she kept crying, her sobs shaking the clothesline on which she hung.

Bones tossed the tire pump aside and turned toward Fat's tree, his anger mounting as it never had before. He would catch that ugly varmint of a fairy even if he killed himself in the process.

While Bones rubbed his empty belly and thought up ways to murder Fat, Fat soaked in a bubble bath. He had spent the night immersed in warm rainwater, plucking bits of hardened stew from his skin and wings. Every joint hurt. He had known he was getting old, but this morning he felt it keenly. Old and fat. He had drunk too many mugs of acorn juice in his youth, and he was paying for it now.

A spasm in his stomach made him double over. He coughed and sputtered as the pain ebbed away. He was no quitter, though. He was no lightweight. "With war comes sacrifice," he said. He pounded his chest with his fists and stretched his wings so that rainwater sprayed everywhere.

※ ※ ※

As if by mutual agreement, Fat and Bones came face-to-face in the middle of the field. Bones had been racing toward the tree with a mind to stuff burning leaves in Fat's hole. Fat had been flying as fast as he could toward the farmhouse, intent on contaminating the drinking water with Bluebell Blindness Inducer. Each man halted when he realized his enemy stood or, in Fat's case, hovered in front of him. Being in such close proximity for the first time unnerved them both. Fat and Bones grew bashful. Bones shuffled his feet and even whistled a note. Fat flew in brief thrusts like a hummingbird.

Bones thought he should make conversation. "My Ma's flat," he said.

Not wanting to seem rude, Fat said, "Yes. She was supposed to be you. Or you were supposed to be her. Be flat, I mean." Poised on the tip of a wheat stalk, he added, "There are marks on my tree."

"Yeah," said Bones. "Cat kept me from chopping it down."

Things might have gone differently from that moment on if a strong wind had not lifted Fat from the wheat stalk and hurtled him into Bones's face. There might have been peace. But broken noses tend not to make friends of enemies, so while both Bones and Fat dabbed at the blood trickling down from their nostrils, the outcome became clear—they would fight to the death.

"I will shred your wings to pieces with a very tiny pitchfork!" screamed Bones.

"If you had a very tiny pitchfork, I would shove it up your big nose into your very tiny brain!" screeched Fat.

"Vermin!"

"Lummox!"

"Pip-squeak!"

"Meathead!"

Fat sliced through the air, shrieking toward Bones with the speed of a young fairy. He rammed into Bones's head, denting the man's skull. But as Fat tried to back up, to fly away, Bones's bramble hair snagged the fairy's wings. Thus bound, the two of them reeled across the field.

"Lemmego!" cried Fat.

"C-can't see," gagged Bones. The dent in his head had damaged his sight.

Running and stumbling, spinning and falling, the two enemies crashed through the wheat, stomping down the slim golden reeds until not one strand remained. The field was destroyed.

From the clotheslines, Mrs. Bald continued to weep. The sun could not dry her eyes fast enough to keep her sorrow from flowing.

Exhausted, Bones finally fell to his knees before his hanging mother. His eyes crossed in confusion. Then he tipped over, crashing to the ground with his hands by his side. The force of the fall freed Fat's wings, and the fairy stood up, pumping his short arms in the air as if he had conquered the world.

The cat, licking up what it could of Mrs. Bald's tears, turned toward him . . .

Before she flattened
like one of the pancakes
her son loved so well, Mrs.
Bald needed one special
ingredient in order to make
her famous pig foot stew.

The Dance

It was almost showtime.

In another few minutes, the sun would descend low enough to transform the field of wheat into a brilliant golden backdrop, low enough to turn the top rail of the peeling fence into a stage, just low enough to lend a little glamour to the pen.

Apple was the star of the show. She hopped up and down in the corner of her pen, loosening up. She needed to be limber, to be focused. She needed to meet the others' expectations, their hunger to see beauty within their bleak surroundings. She would not let them down.

Not today.

Not ever.

This was her dance.

This was their dance.

In the opposite corner of the pen, deep in the shadows, lay Esmeralda.

Esmeralda did not possess Apple's shiny hide or dexterity. She did not even possess four feet. She had four ankles but only one foot among them. Like the others in the pen, Esmeralda hobbled through the mud, pushing herself to the trough and back again. And like the others, she took loving care of her remaining foot, even decorating it with a bracelet of daisies. The flowers grew on the other side of the fence. She had stuck her snout between the railings to gather them, then wove them together using her teeth and spare ankles. This one remaining foot, garnished with white buds, was her pride and joy.

Of all the pigs in the pen, only sweet, young Apple had four feet.

Four lovely feet.

Four pink, supple, enviable feet.

Esmeralda wanted those feet.

She at least didn't want Apple to have those feet. She wanted so badly for Apple not to have those feet that her mouth had filled with bitterness and her head had begun to ache. It had ached for weeks now.

Lucky for Esmeralda, not only was it nearly showtime—it was nearly dinnertime. And the dead farmer's son would be wanting his pig foot stew.

The musicians and singers began prepping for the performance. Neither troupe had very many members, but they had gathered tonight because they wanted to see Apple dance and because pen life provides limited options for social activities. Generally, what one pig does, the others do too.

The chorus consisted only of baritones. They grunted and cleared their throats, warming up by reciting, "A perfect pig is pink and spry," until their lips tingled.

The musicians tuned up, harnessing their intestinal gases. Some time ago, when more of them had sported full sets of feet, the pen had boasted a robust percussion section, but those days were gone. They maintained the wind section thanks mostly to the discount slop that sloshed in the trough.

The sun began to sink into the horizon.

The wheat glowed.

The musicians and the singers were ready.

Apple took her place, her body silhouetted against the wheat backdrop.

"This is going to be the best show ever," said Jeremiah, settling himself on the ground. Until the farmer's wife had taken Jeremiah's two front feet, he had been the best drummer in the pen. He used to bang upon the trough with true ferocity. Now, the music had gone out of him. He used his two back feet, wrapped in ropes of long grass, only to maneuver his way through the mud. But he still loved to watch Apple dance.

"Oh yeah," said Esmeralda. "The best show ever."

She lunged out of the shadows and sank into the mud along the railing, nuzzling her bracelet so that the daisies perked up. She peered through the wooden slats toward the farmhouse. So far, nothing.

All in good time, she thought. She smiled.

Jeremiah smiled back. "Such a beautiful evening. And Apple, isn't she a jewel?"

"A real jewel," said Esmeralda.

"The way she dances, it just, I don't know, it makes me feel—" Jeremiah choked up.

Esmeralda felt she might throw up. "I know," she said. "Me too."

A hush fell over the crowd.

The baritones began in unison with a low sustained note.

The winds chimed in, short bursts first, followed by soft breezy tones.

And then, Apple began to dance.

Apple knew how to move.

When she arched her back, she looked as slender as a half-moon. When she kicked up her feet, she softened the hearts of her audience, made them think of when they were just piglets, fully limbed and free from the cares of the pen.

She shimmied and she shook.

She pirouetted and she tapped.

She lifted her nose to the sky, and the sun seemed to balance upon it, to kiss her, to bless her.

Jeremiah sighed.

Those around him sighed.

Esmeralda spat and looked again toward the farmhouse.

Something was happening. The dead farmer's son barreled out the door with an axe in his hand. His face was red and ugly. With enormous strides, he stalked off into the field—toward the fairy's tree, not toward the pen.

Esmeralda almost despaired, but shortly thereafter, the grey-haired mother emerged from the house with an apron tied around her waist, brandishing the meat cleaver in her hand.

The cleaver. At last!

Esmeralda looked up at Apple, twirling on the fence, twirling on those four lovely feet, and she knew that justice was about to be served.

The winds tightened their tummies to emit a dramatic crescendo.

The chorus harmonized, some notes clashing at first, then slurring into balance.

Apple danced on, a vision of beauty bending to the will of the music.

The woman came closer. She had treaded the route to the pen so often she could have walked it in her sleep. The meat cleaver caught the light of the setting sun and sent its glare across the audience.

Someone gasped, "The cleaver!"

Someone else shouted, "Dinnertime! Run!"

The wind section scattered in all directions.

The chorus cut off abruptly, leaving a note to linger in the air.

Old Flossie, whose two feet let her walk only in a circle, spun around and around until Jeremiah escorted her away from the gate.

Only Apple did not notice. She continued to dance, unaccompanied, as the woman swung the gate open.

"Someone must tell Apple," said Jeremiah, panic on his face.

"You'll never make it in time," said Esmeralda.

Jeremiah looked at the woman approaching. He looked

at Apple. If he were to cross the pen, he would put himself at risk. But Apple! Who else would dance for them? Jeremiah twisted around to look at his feet. Could he afford to lose another?

Jeremiah had the chance to be a hero.

He turned around and fled.

Esmeralda laughed softly, a flash of wickedness in her eyes.

The woman wept and was blinded by her tears. She held out her free hand to feel the way, the cleaver raised in the other hand. She grabbed onto the railing and shoved her feet through the mud. Only a turned corner and a half-length of railing lay between her and Apple.

Esmeralda could scarcely contain her joy. Behind her, the other pigs cowered in the shadows, attempting to hide their ornamented feet. They no longer cared about Apple, no longer cared about the dance.

"Here, piggy-piggy," said the woman between sobs. "I just need one little footsie. One little footsie is all I need."

Cowering behind Old Flossie, Jeremiah called out, "Apple! Apple! Look out!"

But Apple was in reverie.

She dipped.

She spun.

Her tail twitched, and her nostrils flared.

"Don't worry," said Esmeralda. "I'll take care of her."

"Oh, please do," Jeremiah said.

Esmeralda lumbered forward while the woman

approached Apple from the opposite end of the pen.

"Piggy-piggy, come here, piggy," said the woman. She turned the corner. Another ten steps and her hand would find Apple.

"Apple," said Esmeralda softly.

Apple did not hear.

Esmeralda shrugged. What could she do? She had tried. She turned to watch the show—the real show, not Apple's ridiculous tipping and dipping.

This was what she wanted: a nice close-up view. She had been waiting a long time. Already the pain in her head receded.

The woman came closer, extending her legs to secure her footing, her dress drenched in tears.

In her hand, the cleaver whispered, "Blood blood blood, all I want is blood, fresh juicy piggy blood, blood blood blood."

Esmeralda had heard that whisper three times before. It sent chills through her now.

But Apple did not hear the cleaver. She continued to dance.

From the other pigs, her beloved audience, she received no help at all. They, too, heard the call of the cleaver and felt their remaining feet turn cold.

The woman was just three steps away.

The cleaver called out.

The sun descended.

Two steps.

The sky turned red.

The air cooled.

One step.

A member of the wind section could not contain his fear and released a long high note. Old Flossie whimpered.

Apple pivoted, and the woman's hand crept upon her foot, the fingers circling her tender ankle.

Esmeralda grinned, her heart nearly bursting as the cleaver got ready to do its work on one of those four perfect feet.

"Blood!" shrieked the cleaver.

Apple did not feel the woman's hand upon her foot.

Or if she did, she did not care. She had a higher calling to attend to.

Below her, Esmeralda could not believe her eyes. She could not believe that with a human hand upon Apple's foot and the cleaver calling out, Apple continued to dance. Was she insane? What was the meaning of this?

Indeed, Apple spun in the woman's grasp as if there were no hand upon her at all. She twirled and twirled and twirled, ignoring the spiny white fingers.

The wind stirred up a rhythm of its own, and Apple followed it. She lifted her hands to the sky. She lifted her face to the sky.

And then, somehow, Esmeralda understood.

She understood with all of her small heart. Something bigger prevailed. She didn't know what it was, but it was bigger than her, than Apple, bigger than the whole pigpen.

It stretched up into the sky and back down to the earth and wrapped everything in between in what she could only compare to a warm coating of good, black mud.

And she knew she couldn't let the cleaver find Apple.

Esmeralda looked behind her. Jeremiah hid his face. Old Flossie had fainted dead away. The others watched with their mouths hanging open.

Pigs are not accustomed to sacrificing themselves for others. They are accustomed to being sacrificed to the dinner table, but they do not go out of their way to make each other's lives easier.

What Esmeralda did next went down in pig history.

She lunged forward and knocked down the weeping woman. Then, she placed her foot, her last foot, adorned so lovingly with daisies, upon the woman's chest.

The cleaver rose high and called out, "Blood blood blood, all I want is blood, fresh juicy piggy blood, blood blood blood!"

The bracelet of daisies fell to the mud. The woman crushed them with her thick boot when she stood up, Esmeralda's foot in her hand.

Lying immobile on the ground, Esmeralda looked to the railing. Her head had cleared. She tasted only sweetness now, for she had a first-rate view of Apple, dazzling in the sunset, taking her bow.

while Bones slept like a
baby even on an empty
stomach, and Mrs. Bald lay
forgotten on the kitchen
floor, drowning in her own
tears, Fat mixed a potion
in his hole in the tree and
looked forward to soaking
in a bubbly bath. And above
him, in a smaller, cozier, more
charmingly decorated hole,
lived a very sophisticated,
very lonely spider. A spider
who decided that tonight he
would try something new.

Leonard Grey III

Leonard Grey III set down his pencil.
He had been trying to compose a sonnet.

He sighed. He couldn't concentrate.
Leonard mourned the dead man, who,
in a fit of anger, had years ago thrown
a pocket-sized book of sonnets, an
unwanted gift, out the kitchen window.

It had taken Leonard days to hoist
the book up to his hole in the tree,
where he had read it every day since,
finding solace in the beautiful language.

So what if he wasn't good at sneaking about like a good spider should? So what if his family mocked him for his clumsiness and his love of tea? So what if loneliness sat upon him like a rock? He had his poetry.

But that night, not even poetry could ease Leonard's loneliness.

Leonard sat in his doorway, staring at the blue-black sky beyond the heavy leaves of the tree.

Below, a spider hung strung up in the fairy's den. A cicada had warned Leonard earlier that tonight Fat was brewing a batch of Bluebell Blindness Inducer, which required a large quantity of spider blood, enough to drain a spider Leonard's size to an inch of his life. And Leonard was no mite; he stood nearly as tall as Fat himself.

Fat was holding another spider captive. Fat was going to drain that spider of its blood. And if Leonard could only learn to sneak, he intended to rescue that spider. Then maybe, just maybe, that spider would have a cup of tea with him.

"It's going to work," Leonard said aloud, "I'm a spider, that's what I am, and sneaky is my middle name." He leaned too far out as he spoke and slipped.

"Dagnabbit!" He scrambled back into his hole. He drew in a foot to cover his racing heart. "Breathe, Leonard, breathe."

He went inside and put the kettle on.

He thought for a moment, deciding that before he attempted sneaking, he needed proper gear. He donned a stocking cap, boots, and sunglasses to conceal the glow of his eight eyes. Secrecy was vital.

"I can do this. I am a spider." He said it in his deep poetry-reading voice, but he couldn't convince himself.

Even so, stubbornly, he set forth.

As he began a careful—*whoops!* He had forgotten about that bent piece of bark—descent toward Fat's hole, he wondered what he and his new friend would talk about. Philosophy, maybe? Politics? Perhaps his new friend was a traveler and would tell him stories about other places in the world!

Thunk. He tripped and fell upon an obliging leaf some four feet below. If the fairy heard him coming, he'd stand no chance of executing the rescue mission.

His heart pounding, he raised and lowered his legs, settling each one back down on the bark. Though normally Leonard could see fine in the night, the sunglasses made everything blurry and dark. Leaves loomed all around him, rustling and settling in for the night.

He neared the fairy's hole. He flattened himself against the branch and took deep breaths. So far, he had not caused too much commotion, at least not enough to alert Fat of his approach.

He heard Fat inside the hole, moving vials of potions about. The fairy was muttering the recipe for Bluebell Blindness Inducer. "Six bluebell petals. Three rabbit droppings. One cat hair. One human toenail clipping; thank you, Bald, may you rest in peace. Eight drops of midnight dew. And finally, a generous quantity of fresh, warm spider blood."

Leonard held his breath. Had he arrived too late? Had his new, yet-unknown friend already withered and died? Trusting his camouflage gear to conceal him, Leonard peeked into the hole.

To the left, the fairy stood at his worktable with his back to the room, humming a song as he measured and mixed. His wings were threadbare, but his arms showed his strength. He was old but no weakling. On the table, beside a large beaker of blue liquid, lay a large knife with an intricately carved wooden handle.

To Leonard's right, a spider hung in the corner, legs tied together. He was either unconscious or asleep or dead; Leonard couldn't determine which.

It seemed too easy. With the fairy engrossed in his work, his back to the room, all Leonard had to do was sneak in, set the captive free, and the two of them could hightail it out of there, back up the tree.

If Fat did turn around and see him, hopefully it would be after Leonard had set free the spider. Then the odds would be two against one. And Leonard had a feeling even Fat would hesitate to take on two spiders, even if one of them was a coward.

Settling his stocking cap a little lower on his head, Leonard almost fainted from panic before silently scolding himself to get moving.

He ran into the room before he could talk himself out of it.

After a few steps he tripped, performing something like a somersault as he pulled out of the fall. He was almost there!

And then, quick as lightning, he was thrown on his back, his boots waving in the air. His sunglasses fell off, and a foot landed upon them, a wrinkly foot that smelled like onions, squashing the dark lenses and twisting the rims.

"Hello, neighbor. Pray, tell me what your hairy, lumbering self is doing in my hole," said Fat, the long knife in his hands.

Stunned, Leonard could only gurgle in response.

"You are a terrible excuse for a spider," Fat said, rolling his eyes, "I heard you when you were halfway down the tree. You might as well have hired a blue jay to announce your arrival." He twisted the knife in his fingers. "Pathetic."

Leonard did indeed feel pathetic, splayed out on the

ground, unsure of how he got there. The indignity was extreme.

A thin whistling sound came from outside.

He had forgotten the kettle!

Fat groaned. "Do you never tire of drinking tea?"

Leonard gurgled again and struggled to right himself, but Fat placed his foot upon Leonard and held him there.

"Well, I tire of hearing your kettle go off, that's for sure," Fat said.

Fat placed the knife's tip at Leonard's abdomen. He clearly enjoyed toying with his second captive.

Leonard was at a loss.

With his feet wriggling in the air like some infant's, his terror grew so great that he couldn't swallow. Spit began to collect in the corners of his mouth.

He was not a normal spider, so why had he tried to be? His plan could only ever have ended badly. He should have known he would bungle a rescue attempt. He was a failure at spiderhood; he should have admitted as much. Furthermore, the spider in the corner had not roused. Leonard began to doubt whether the poor creature had any life left in him at all.

Leonard knew he would die today; he would die this moment.

And would that be so bad? He had nothing to lose. No one would mourn him. And that was sadder than actually dying. Leonard wanted to be missed.

I'm not ready to die, he decided. I need to find someone to love.

And once he figured that out, he got sort of brave.

Shaking, he pushed aside the knife.

"No," he said.

"What?" said Fat, staring at the knife in surprise.

"No," said Leonard. He thrust his feet upward in an attempt to right himself—not a graceful movement but an effective one. He stood facing the fairy.

"No?" said Fat.

"That's what I said."

"No what?" said Fat.

"Just no," said Leonard.

"You can't say, 'just no,'" said Fat. "I have a knife." He held the weapon up again.

Leonard sensed Fat's annoyance, his confusion, and homed in on them. He knew what he needed to do. He might have been a lousy spider, but he was an above average poet.

Fairy, you winged patutem, spangly voo,
Thy shredded heart, like vaporish splantshine
Does unto earth like the beast of the moo
The ungreen plate of your bonderoo dine
A cud and a cud and a cud and a cud.

Fat spat on the floor. "Are you trying to recite poetry? You're terrible! And there's no such word as *splantshine.*"

He pursed his lips. "Though you do have a knack for rhyme."

Leonard continued:

Fast like the wind on a sperry pink flate
Your dank blackened mubgaw stinks like the fud
You hate the stars, you hate the sun, you hate—
But hate—it won't win—for love came here first
To the splight ball of blue, spinning in space
This giant tiny thinker dink of thirst
On which we all spin at a vash-bash pace.
Here, know, that no skander-winged puck, shall live
Apart from the holy call to—

With stealth of a kind that Leonard could never match, the captive spider slid behind Fat, gazing at Leonard over Fat's shoulder with the most beautiful green eyes Leonard had ever seen. Here was not a male spider, as Leonard had supposed. No, here was an Aphrodite.

Leonard was tongue-tied.

Unaware that his former captive stood a breath behind him, Fat tapped his brow with the knife. "How about *convive?* No, that doesn't quite work, does it." He scrunched his mouth and studied the ceiling. "Apart from the holy call to . . . *active?*"

The other spider, whose name was Priscilla Mae, tapped Fat on the shoulder.

Fat spun around, but Priscilla Mae sidestepped behind

him. She had snuck out the door, into the night, before Fat had time to register the empty corner of the room.

"Unbelievable!" shouted Fat, turning back to Leonard. "You spiders sure stick together, don't you? One spouts god-awful poetry while the other breaks free."

Leonard was pleased beyond measure by the turns of events. He smiled uncontrollably.

Fat moved quickly, brandishing the knife. "You owe me blood, arachnid." He drove the blade deep into Leonard's front right leg.

Leonard staggered back from pain, crumpling to the floor.

Fat fetched a beaker and set it beneath the wound, patting down Leonard's leg hairs so they wouldn't disrupt the flow.

"Very good," Fat murmured after some time. "A most excellent burgundy. Really excellent. And not a bad consistency, either."

Leonard began to feel lightheaded. "Excellent."

"That's what I said," said Fat.

"Excellent."

"Stop repeating what I say."

"Stop repeating what I say," said Leonard. He had forgotten where he was.

Fat raised the knife. "Do you want me to give that wound a twin?"

"Excellent."

Fat peered across Leonard's eight eyes. "Hmm," he

said, "I might have drawn too much." He shrugged.

Higher in the tree, several limbs above, the kettle shrieked inside Leonard's hole.

"My kettle," whispered Leonard.

Fat kicked aside the broken sunglasses on his way to his worktable. He poured Leonard's blood into the Bluebell Blindness Inducer potion and swirled the mixture.

The kettle's scream cut through the night.

"Where can a fellow go for a little peace and quiet?" Fat shouted, banging the potion on the table.

"Kettle," whispered Leonard.

After fanning his wings and stomping his feet, Fat realized that if he did not want the tree to burn down, he'd have to take the kettle off the fire himself. He glared at Leonard, slumped in the corner. The spider had ruined the evening.

And yet . . .

The spider was also his neighbor. Did rules about helping a neighbor in need still apply in this day and age? Fat sighed and threw the knife onto the worktable.

"I am ever one for sacrifices," he said and pulled out a clean white cloth from the cupboard. He fluttered over to Leonard and placed the cloth upon the spider's wounded leg, staunching the flow.

"You sit tight," said Fat, straightening Leonard's stocking cap. "I will return shortly."

While Fat flew out to tend to the kettle, Leonard slowly regained consciousness. He could hardly move his legs, but his eyes cleared a little. By the time Fat returned, kettle in hand, Leonard had dragged himself to the middle of the room and propped himself against a chair.

"You're recovering well," Fat said. "That's quite a constitution you have." He held up the kettle. "I saved the tree from burning. Shall we share a cup to celebrate? I think I have an herbal variety around here somewhere."

Leonard did not want tea.

He did not want to stay and chat.

However, being unable to move, he had no choice but to curl a leg around the teacup Fat set on the floor.

Tea made Fat convivial. "Tell me, neighbor, how did you come by your love of poetry? You see, when I was young, I kept a journal of verse. I was no master of the form, you understand, but I came up with one or two pretty turns of phrase."

Fat droned on. To hear the fairy talk, a listener would think he had been the one starved for conversation.

But Leonard didn't hear or care. Weak though he was, he smiled.

Before Priscilla Mae had slipped into the night, she had whispered to him, "Find me under the eaves. Poetry moves me."

A single whiff of love had swept Leonard free of gloom.

"What a day, eh, neighbor?" said Fat. He looked around his hole with a satisfied grin. "Incidentally, how did that other spider escape?" He chuckled. "Now she was a sneaky one."

In the deep, dark night,
unpopular but necessary
creatures emerge from their
hiding places and undertake
unpopular but necessary
activities. When dawn shoves
the night away, no one speaks
of these things, including me.
But dawn has not arrived.

The Dead Man Song

Priscilla Mae crept back into the night,
away from the fat fairy's tree.

From a distance, she heard singing.

She paused, dangling in the cool
night air.

A thousand voices rose, as clear as
the stars above.

Priscilla Mae listened carefully,
eagerly, to discern the words of the song.

Dead man, dead man, once a beast,
Now a splendid midnight feast,
Newly dead, your blood still fresh,
Best of all's the softest flesh,
Dead man, dead man, once a beast,
Now a splendid midnight feast.

Maggots—she should have known. Not a pretty sight, seeing maggots do their work, but they did their job awfully well. And anyway, it wasn't as if that old farmer deserved better. He really had been a beast.

Famished, Priscilla Mae carried on her way, the song stuck in her head.

Mrs. Bald lies on the floor weeping. Her heart is breaking. My own pain is salved somewhat by the happy plan of revenge inside my brain. But it's not quite time for revenge—it's time for love. Not to worry, this is not a love story with singing cherubs and heart-shaped chocolates. This is a love story with a wicked turn.

Jimmy in Love

Alice. Oh, Alice.

From dawn until dusk, Jimmy thought of her. At night, he dreamt of her. His head, his heart, filled with the abiding love of which poets write.

That is a whole lot of love for so small a mouse.

Alice, meanwhile, was the biggest mouse on the farm.

Not that Jimmy cared a whit whether Alice was as large as an orange or as small as a walnut. He loved her, and that was that.

When his father said, "Still pining after that Alice, Jimmy, my boy?" Jimmy would say, "You betcha."

When his brothers teased him and said, "You gonna marry that big mouse and let her boss you around?" Jimmy would say, "You betcha."

When his mother said, "Jimmy, are you going to actually talk to Alice one day?" Jimmy would say, "You betcha."

But one major obstacle had prevented Jimmy from talking to Alice. He dreamt and loved from a distance, for an ocean of kitchen floor lay between them. And that floor was ruled by the very large, very unpredictable cat.

On some occasions, the cat would permit the house's mice to travel across the floor unbothered. He was even rumored to have once said pleasantly, "Good afternoon, mousie-pie." On other occasions, he would rest on his side, tail flipping nonchalantly, then spring up and crush a mouse between his teeth.

A mouse never knew what to expect from the cat. Not even Jimmy wanted to risk such fickleness, even to speak with his beloved Alice.

One day, however, Jimmy overheard Smitty talking to Buster.

"Alice," said Smitty.

"Lordy, you aim high," said Buster.

"She's the one for me," said Smitty.

"When will you ask her?"

"Tonight."

"The cat?"

"I'm not afraid."

Jimmy did not need to hear any more of the conversation. Smitty was a handsome mouse, arguably the handsomest on the farm. A fire started in Jimmy's heart. It traveled up into his head and down to the tip of his tail. He knew he could waste no more time.

At dusk, after finishing his chores, Jimmy made plans to cross the kitchen floor. If he lost his life in the pursuit of Alice, then so be it. Many mice had perished for less noble a cause than love.

"Don't do anything stupid, dear," said Jimmy's mother.

"Give that Alice the kiss of her life," said Jimmy's brothers.

"Come back in one piece, now, you hear?" said Jimmy's father.

"You betcha," said Jimmy.

Leaving behind his family's cautions, Jimmy stepped out into the darkness and stood against the baseboard of the second-floor hallway, straining to see out over the kitchen. He took note of every small movement. His ears stood straight out, picking up every sound.

The farmhouse seemed quieter than usual.

Night had fallen fast.

To Jimmy's right, stairs led down to the kitchen. He scurried to them and jumped from one to the next, stopping to listen for some sign of danger.

Nothing.

He jumped to the next step.

Still nothing.

He jumped down a few more.

Jimmy inhaled deeply—it was now or never—and then took the remaining steps in consecutive leaps, skidding to a halt at the bottom.

The kitchen was dark except for a thin beam of moonlight shining through the curtained windows. Something shiny and reflective covered the floor. Jimmy stretched out his foot and pulled back in shock.

Water.

Jimmy had no way of knowing how deep the water was, and he did not wish to jump in, both because he didn't know how to swim and because before he left, he had licked his coat until it appeared as smooth as velvet. If he got wet, he would look like a ragamuffin and stand no chance of competing with that Smitty.

Oh, that cursed Smitty. That handsome, cursed Smitty. Smitty would not let a wet floor stop him. Perhaps Smitty was on the other side of the kitchen already, whispering plans about a dream nest into Alice's ear. Perhaps he was tickling her big, beautiful paws and telling her what he intended to name their many children.

Jimmy ground his teeth.

Think, Jimmy. Think.

He needed a boat.

With some effort, he jumped back up the stairs, panting when he reached the top. He ran back into the bright family nest and threw the scraps of cloth out of the measuring cup that served as his bed. Then he grabbed a metal souvenir spoon with a windmill on the handle and lugged it inside the cup.

"Everything all right, Jimmy?" said his mother, coming out of her bedroom in her nightcap.

"You betcha," said Jimmy.

He hauled the measuring cup out of the nest, into the hallway, and carefully dropped it down each step until once again he stood at the bottom, water in front of him and Alice somewhere far away to the left.

Every second he lost was a second Smitty gained, so he jumped into the measuring cup, dipped the spoon into the water, and began paddling vigorously.

Left, right. Left, right. Left, right.

He used his tail as a rudder, directing himself across

the floor. He paddled for he didn't know how long. He stopped, just for a second, just to let his weary shoulders relax, when he heard a low wheezing sound.

The cat!

Back he sprang, brandishing the spoon like a sword.

Wheeze.

Jimmy's heart raced, his mind raced, his blood raced, and he wished he were racing right back up the stairs.

He would have, too, had it not been for the lingering vision of that Smitty tickling Alice's feet. Smitty would not have let the cat stop him. In fact, Smitty was probably with Alice right that moment.

Right that moment!

Wheeze.

The measuring cup came to an abrupt stop. Instinctively, Jimmy ducked. He cowered on the floor of his small craft, imagining a giant paw swinging down from above, crushing him completely.

And still, he thought of Alice.

He cautiously stood up and then lifted one paw out of the measuring cup, feeling around for whatever he had bumped into.

His toes brushed against something—a little mushy, yes, and damp, but relatively firm. Not cat but land!

He jumped out of the measuring cup, still holding onto the spoon.

Wheeze.

The sound came from below his feet.

Jimmy kneeled down. A thin line of something grasslike lifted slowly before him, revealing an enormous weeping eye.

Wheeze.

The eye blinked.

Sniff.

A sniff of a different pitch entirely.

Jimmy whirled around, and his nose bumped into something furry.

"Hello, mousie-pie."

Jimmy's life flashed before his eyes. Or, rather, the life he would never live. He saw Alice, he saw their nest, he saw their children—he saw all that would never be. He saw his life given away to that Smitty.

"Going someplace?" said the cat.

"You-you-betcha," said Jimmy. He took a step backward and landed on the eye.

"Get off her eye, rodent!" hissed the cat.

Jimmy stumbled. He saw now that there was a second eye on the floor. Both blinked. Both wept.

Jimmy turned to hop into the measuring cup and paddle for freedom, but a paw on his tail held him back.

"Where are you running off to, mousie-pie?"

Jimmy was too terrified to answer.

The cat stuck his nose in Jimmy's face. "I'd like to get your thoughts on this," said the cat. "Can I count on you to give me your honest opinion?"

Jimmy had no choice but to nod.

"Good." The cat slumped his shoulders and then swerved to show Jimmy his rear end. A shaft of moonlight revealed the cat's mangled tail. "Do you think it will grow back?" he said.

The fur was clotted with blood. Jimmy could see the bone. The tail did not look like it would grow back, but Jimmy said, "You betcha."

The cat sniffed. "Really?"

"You betcha."

"Oh, I always knew you mice were a decent sort of creature," said the cat. "Beyond decent tasting, I mean. You're a sensitive lot."

Wheeze.

The cat shifted on his haunches. "Oh dear, oh dear," he said. "She just won't stop crying." He licked the tears from the weeping eyes.

A large nose south of the eyes discharged another wet, wheezing noise.

Jimmy backed up against the edge of the cup. The spoon banged against the glass.

"Where are you going, mousie-pie?" The moon shone on the cat's face, on his narrowed green eyes, on his pink protruding tongue, on his white teeth.

Jimmy didn't answer. He didn't hesitate. He didn't want to stick around on that wheezing, weeping flat face any longer. He jumped into the measuring cup and took off, working the spoon left, right, left, right, as if his life depended on it. Which it did.

The cat's rising paw blocked out the moonbeam. Jimmy ducked, feeling the wind above his head.

The cat's hisses sent waves across the water, driving the measuring cup forward, threatening to capsize it.

Soon darkness overtook the kitchen.

Jimmy heard, *wheeze, sniff*—then, "Oh mousie-pie, oh mousie-pie, sweet little mousie-pie." But from a distance now. The voice grew faint.

Exhausted, Jimmy collapsed at the bottom of the measuring cup. For the first time in his life, he wished he were bigger, stronger, hairier, and more threatening. He should have shoved the souvenir spoon into the cat's gaping wound.

He should have . . . he should have . . .

Cold, alone, and miserable, Jimmy realized that the reason he had never spoken to Alice was because he had done so little in his life. And thus he had nothing to say.

Nothing but "you betcha."

And what girl wanted to hear that all the time? Those were not the words of love. Those were the words of a small mouse. An insignificant mouse. A mouse that—

"Ahoy there!"

Jimmy's head shot up. He had reached the other side of the kitchen, but not in time.

In the shadowy darkness, Smitty stood on top of the stove. And next to him, big, beautiful Alice.

Oh, that cursed Smitty! That handsome, cursed Smitty!

Jimmy sprang to his feet. He threw his fists into the

air. He opened his mouth to finally curse Smitty to his face when the measuring cup flipped over, throwing Jimmy out into the open water.

He surfaced just in time to see Smitty leap off the stove and perform a flawless swan dive.

Smitty threw his arm around Jimmy's neck and dragged him to the side of the stove, where a cupboard door stood open.

Alice stood in the cupboard doorway, reaching out her hand.

"Got him?" said Smitty.

"I sure do!" said Alice, hauling Jimmy inside with ease.

Bedraggled and half drowned, Jimmy stared at the two faces studying him.

"His name's Jimmy," said Smitty.

"Yes, I know," said Alice.

She knew his name. Her voice was rich and deep. It sent shivers through him.

"He's cold," said Alice. "Smitty, go fetch him a blanket."

"Right away," said Smitty.

"Oh, you poor dear," said Alice, shaking Jimmy's whiskers free of water.

Jimmy gazed upon his beloved and opened his mouth.

This was his chance. It was now or never. Smitty was gone. And Jimmy was alone with Alice.

He had seen the weeping eyes, fled the capricious cat, and had braved the kitchen floor for this very opportunity. He must speak. He must tell her how much he loved her.

He practically worshipped her. He wanted to marry her. He wanted to ask her if she'd be all right naming their first son Jimmy Jr. He'd always wanted to have a son named after him.

"Uh," said Jimmy.

"Yes?" said Alice.

"Uh—"

Alice leaned in, concern on her face.

"Uh—"

If only Jimmy had spoken.

If only.

Smitty returned with the blanket, unfolded it, and wrapped it around Jimmy's shoulders.

Jimmy nodded his thanks.

Smitty stood up. He took Alice's hand in his own. His handsome face glowed with pleasure and pride. "Jimmy, old boy, I guess you'll be the first to know the good news." He inhaled and said, "Alice and I are getting married."

Jimmy's body went numb beneath the warmth of the blanket.

Alice blushed and ducked her head and could not have looked lovelier. "You'll come to the wedding, won't you?" said Alice.

Jimmy wanted to die. He wanted to die right then and there. He tried to shake his head but could not say no to Alice, so he nodded instead.

Smitty said, "I can't wait to tell my folks. Can we borrow your cup and spoon to get back across? I came over on a graham cracker, but I'm sure it's dissolved by now."

Again, Jimmy nodded against his wishes.

"Thanks, Jimmy," said Smitty, "you're a real sport." He jumped into the measuring cup and extended his paw to Alice. "Hop on in, my love!"

Alice smiled and smoothed her tail. She sashayed to the measuring cup and stepped in.

"Oh," said Smitty, as the cup sank dangerously low.

"Oh," said Alice, as the cup rocked back and forth. She hopped out.

"I think—maybe . . . ," said Smitty.

"Yes," said Alice. "I think perhaps I should wait behind."

"No matter," said Smitty brightly. "I'll be back in a jiff with a larger vessel." He lifted the spoon. "Look after my bride-to-be for me, will you, Jimmy old boy?"

Jimmy's heart stopped. He looked at Smitty. He looked at Alice. A thunderclap sounded in his chest.

He loved.

He hated.

He put his arm around Alice, and he said, "You betcha, Smitty, old boy. It's a long trip back. Make sure to take a break on the bit of land over by the moonbeam."

The sun has risen. The
full moon has retreated.
But death persists. Pollen,
such a deadly weapon.

Petals to a Pollen Gun

Tulip and Daisy were best friends. They had been friends ever since they were tiny green sprouts, reaching for the open sky on the edge of the wheat field.

Daisy was friendlier than Tulip, but Tulip was more intelligent. Because each flower had blossomed in her own special way, they did not feel the need to compete with each other. In fact, they agreed that they would rather die than allow anything to come between them.

In their straightness and strength, their richness of color, and their gift for brightening the faces of passersby, they were the epitome of floral beauty.

They grew along the east bank of Puddle One.

Puddle One was not large—it was about the size of a student's desktop—but it served the flowers' needs. They never went thirsty.

"Oh, Tulip," said Daisy on a sunny autumn day, "how wonderful it is to be alive."

"Isn't it though?" said Tulip. Her bloodred petals glowed in the sunshine. Not for the first time, she wished someone would paint the pretty picture she made. She and Daisy.

"I just love life," said Daisy, white petals bright in the sun.

"Life is good," agreed Tulip.

"It's not just good. It's divine!" Close by, a songbird echoed Daisy's joy with a trill.

"Divine, then," laughed Tulip. She leaned toward her friend, and they touched petals.

At that moment, on the other side of the puddle, a crack began to form in the earth. The crack was tiny, not more than a pebble long. A green sprout peeked out.

"You are so beautiful," Daisy was saying to Tulip.

"I was just going to say the same about you," said Tulip.

"You are more beautiful," said Daisy.

"If that is true, and it's not, then you are the most lovable," said Tulip.

"Oh, stop."

"You stop."

So devoted were Tulip and Daisy to admiring each other, they did not notice the earth crack again on the other side of the puddle.

Another sprout peered above the ground.

The earth cracked again and again and again. Each time, a sprout peeked out to see if this was a good place to grow. And every sprout saw Tulip and Daisy, strong and beautiful, and Puddle One, presumably the reason why

Tulip and Daisy had become so strong and so beautiful. This, thought the sprouts, is a very good place to grow.

So grow they did, stretching their fledgling roots toward the puddle's other side.

And Tulip and Daisy did not notice.

They did not notice until the moment when Daisy could not draw enough water from the puddle to satisfy her thirst.

"Tulip!" said Daisy. "Puddle One is nearly dry!"

"Impossible," said Tulip.

"Listen," said Daisy, sucking and slurping, creating a hollow garbling sound.

Tulip's roots made the same sound. She managed to draw up only a thin line of water. It was then she saw the row of healthy green sprouts on the other side of the puddle, chattering and gurgling.

"Why, those thieving infants," Tulip said. "They're sucking our puddle dry."

"What!" cried Daisy. "Then what will we drink?"

For the briefest moment, Tulip didn't know what to do.

Daisy looked at her expectantly. Tulip was, after all, the more intelligent flower.

"You must ask them to leave," Tulip said.

"Me?" said Daisy.

"You are the friendlier flower," Tulip said.

"Yes, but—" Daisy said. "Oh, very well."

She formed a bullhorn of her petals and said loudly and clearly, "Excuse me—little friends? This puddle is ours. Please move on to a different puddle, or there will not be enough water for Tulip and me. We were here first, after all. Please understand, we mean no offense." She paused, then added, "And we wish you the utmost joy in your new life somewhere else."

One of the sprouts let out a low, gurgling burp, impossibly loud for so small a plant.

Shocked, Daisy pulled back.

"How rude," said Tulip.

"They sound drunk. Drunk in the way Farmer Bald used to be."

"Drunk on *our* puddle water," Tulip said. "But will they move, is the question."

The sprouts began burping again, as though they were having a contest. Was that giggling? At the burps or at Tulip and Daisy? It was clear, at least, that not a single sprout had pulled under to move elsewhere. The pests were rooted to their spot.

Daisy shook her head. "I don't think they're going anywhere. And without water, soon you and I will begin to wilt, and then our color will fade. And then the stink will come." She startled herself with this last observation. "Tulip, we're going to stink!"

The thought of her bloodred petals fading sickened Tulip. After a moment, the sick feeling turned to anger.

"Enough," said Tulip. "I'll handle it." Her petals already bullhorn-shaped, she hollered, "Keep away from our water, you little squirts!"

"Tulip," gasped Daisy. "Is that kind of language really necessary?"

A burp came from across the puddle.

"This is too much," said Tulip. If she only had herself to worry about, that would be one thing, but she had taken Daisy into her care when Daisy's family had all been picked by that big, mean pig. Now, she and Daisy were sisters, two flowers against the world. They needed only three things in life: sun, dirt, and water. But without even one of these things, well . . . Tulip formed a plan, to be acted upon straight away.

"Daisy," she said. "We need to take action."

"What do you mean, take action?"

"We've asked them to leave. They've refused. Now, we've got to put them in our sights."

Tulip tested her pollen-release capabilities. *Foof.* A small yellow puff whiffed overhead, then drifted lazily down. Not bad for a first try, tighter next time. "We'll shoot them down with our pollen."

Seeing her dear friend speak in such ugly terms, Daisy grew confused. "You mean . . ." She lowered her voice. "Shoot, as in, kill them?"

Tulip popped off another blub of pollen, and it dropped like a miniature bomb into the puddle mud. Next time, she'd aim out, not up.

What a bold new feeling this was! Why had she not thought of this before? She might have saved Daisy's family from the pig. She might have protected their water supply. Nobody would mess with them again. Pretty, passive flowers? Not anymore.

"Tulip," persisted Daisy, "are you suggesting we kill the sprouts?"

Tulip replied, "Please, Daisy, you've got a big heart. That's what I love about you. But don't go having a moral crisis about this, all right?"

Aiming out, not up, Tulip released a round of pollen pellets that riddled the opposite bank. "Did you see that?" she laughed.

Daisy judged the holes to be about four pebbles deep. The sprouts wouldn't stand a chance. "Can't we try talking to them again?"

"Come on," Tulip said. "Try it. Just *shoot*."

"I don't know."

The sprouts' heads bopped about on the other side of the puddle. They were babbling happily, as babies do.

"They're so little," Daisy continued.

"Please. The issue is simple. There's a limited supply of water. It's us or them."

But having already spoken up, Daisy found she could not back down. "But we're flowers, not"—she searched for the word—"assassins."

"It's self-defense."

"You aren't hearing me," Daisy said.

Tulip rolled her head. "Stop thinking so much! I'm the brains of this operation, remember? The beauty too, if you stop to think about it, because white isn't even a color." She looked across the breadth of Daisy's head. "Red, on the other hand . . ." she preened, "Everybody loves red. It's the color of the only thing that matters." She bent to touch Daisy's petals, but Daisy pulled away. "Blood. Spill it, or have it spilled."

Daisy knew then that Tulip's mind was made up.

The wheat surrounding them stirred, suggesting an oncoming storm, yet the sky remained cloudless and blue.

Daisy shivered, but even so, she spoke. "I'm not as smart as you, that's true, but I do know the difference between right and wrong, and this is wrong."

"Suit yourself," said Tulip. She aimed and, in a rapid-fire burst, blew apart a tiny green sprout.

"Gotcha!" shrieked Tulip. Now *this* was flower power!

The other sprouts bopped crazily about, trying to pull under or run away, but none seemed to remember exactly how it was they had arrived there in the first place. They chattered loudly, like a cadre of angry squirrels.

Daisy couldn't stand by and watch, she just couldn't. Acting on instinct, she swung around and whacked Tulip with her head.

"Ow! What was that for?"

"I'm not going to let you kill those sprouts." Daisy

tucked her leaves in front of her. "So if you want to fight, fight *me*."

Rubbing her bruised stem, Tulip said, "It's hardly fair. I'm so much bigger than you."

"Tell that to the sprouts."

Tulip leaned back until her head touched the ground, then sprung forward and walloped Daisy with all her might.

It was like being pounded by a hammer. Daisy hit the dirt hard, her petals crushed and oozing fluid.

Tulip fared better, only two red petals suffered minor tears. "Now you're bald," she said. "How about that."

She retrained her attention on the sprouts, who continued yammering away.

"Shut up!" she shouted. She fired the deadly pollen again and again, her long stem pulsing with every release. Blown to bits, the sprouts formed a bright green streak on the dark earth.

Tulip surveyed the scene. "Well," she huffed, "That about does it. Simpler than I thought it'd be, really."

Lying on the ground, Daisy whispered, "Oh, Tulip, what have you become?"

Daisy hurt all over. She felt naked and ashamed. How had it come to this?

The wheat around the puddle shook softly in the wind, strand after strand rustling, growing. Then the strands of wheat began to fall, stepped on, stomped on.

Close by, Daisy heard Tulip's tired panting and the confused thoughts rolling around deep inside her.

Farther off, she heard the good-blood-love twang of
the bond between mother and son,
    the crying of a foot,
    the munching of flesh,
    clumsy, lonely footsteps,
    the ebb and flow of mucous in a dog's nose,
    bones cracking, minds cracking,
    the earth cracking—
    but winding through it all, like a single golden thread,

a single-note life-song,

a songbird atop the fairy's tree,

singing.

Then—

someone shouted, "Vermin!"

A smaller, shriller voice screamed, "Lummox!"

"Pip-squeak!"

"Meathead!"

A lumbering foot pressed sprout pulp deeper into the ground, and another foot descended upon Tulip's head, her bold and beautiful head. Because Daisy had listened, Daisy ducked, to live to see a better day—

A better, more beautiful day.

Puddle One was hers alone now, to share—if she wished.

Just before Fat and Bones met in the field, face-to-face for the first time, a tired, old hound dog returned to the farm. He was harmless, except for his cold. I've heard it said that a butterfly can flap its fragile wings and cause a tsunami thousands of miles away. Might a dog's innocent sneeze have a similar impact? A dog's sneeze might, say, blow across a field of wheat and send an aging fairy hurtling into the face of a dead farmer's son just when the two are getting to know each other better.

What the Sneeze
Said to the Moon

Dog Alfred sniffed.

Darn this cold, he thought.

He sneezed.

He had been sniffing and sneezing for days now, which was the reason his visit to his mom had ended early.

"Get home," she had told him. "And get well."

"Ma," he said, "I came all this way. I can't go home now."

"You live next door," she said.

Dog Alfred blew his nose, expertly aiming the thread of snot into a stack of hay in the corner of the barn.

His mom gave him a disapproving look. "Days are growing shorter, nights are growing colder. You'd best get well before winter sets in."

The thought of lying in front of a fire was too tempting to ignore. Dog Alfred nuzzled his mom and left.

And came home not to a warm hearth but to a cold grave.

He sniffed the earth and the body it held. He could still identify the man's tangy odor.

Bald had not been a good man, nor had he been a bad man. He had just been a man. All dogs with discerning sniffers know that good men smell musky when they die, bad men smell rotten, and just so-so men smell tangy. Tangy like a lemon that's been chewed on and sucked on and batted around in the dirt for a while.

Dog Alfred hung his head. He didn't feel sad. He felt lost. Bald had not given Dog Alfred love, but he had given the dog direction.

"Follow me, boy," he'd say when they went hunting. Dog Alfred would dutifully follow.

"Fetch."

Dog Alfred would fetch.

"Sit."

Dog Alfred would sit.

"Stay."

Dog Alfred would not move, save for raising his heavy eyelids to see Bald better against the glare of the sun.

Dog Alfred liked being told what to do. It saved him the trouble of deciding what to do on his own. He wasn't like the cat, who never seemed idle. Even while sleeping, the cat appeared to be thinking something or waiting for someone or plotting something against someone.

Where was that cat, anyway? The cat usually trailed Mrs. Bald around as if she were made of sardines.

Where was Mrs. Bald?

And where was that no-good Bones?

When Bones dies, thought Dog Alfred, he'll smell like rotten eggs stirred in spoilt milk with a dash of pee. Really yellow pee. Eggs and milk and really yellow pee, that's what he'll smell like.

Dog Alfred yawned, his eyelids twitching up and down. He wandered around the farm.

"What's going on?" he said to a big, black pig.

"Not much," said the pig, lying on her side on the ground.

Dog Alfred wished to appear purposeful in front

of the pig, this big pig whose eyes shone as if she held a secret. "Nice day for a walk," he said.

"I wouldn't know," said the pig.

Dog Alfred observed that the pig did not have any feet. "Oh, I'm sorry," he said.

The pig smiled. "Don't be," she said.

It must be the worst tragedy in the world to lose one's feet, Dog Alfred thought. "Don't you miss them?" he said, pointing at her stumps with his nose.

"It was worth it," she said.

"Oh," said Dog Alfred. He didn't understand. And his discomfort increased the longer he stared at her stumps. "Well, I'd best get going."

"Enjoy your walk." The pig wiggled her ears.

"Thanks. You enjoy, uh, sitting there."

"Will do."

Dog Alfred shuffled along. He sniffed the earth. He sniffed the air. He sniffed everything he passed. When he had sniffed himself tired, he wandered to the side of the house.

He found Mrs. Bald hanging from the clothesline. He wondered why she was not inside cooking. She was always cooking. Maybe she'd decided to try something new because Bald was underground and everything felt topsy-turvy for her. Dog Alfred knew that if his mother died, he'd probably want to hang from a clothesline too.

The cat sat by Mrs. Bald and gave Dog Alfred a dirty look.

Dog Alfred moved on. His cold had begun to make his head hurt. A sudden thirst overtook him.

He sneezed.

His sneeze, a blithe little thing, blew across the field, where someone shouted.

Dog Alfred lifted his head to see what was happening, but the wheat stood too high.

He gave up and wandered back to Bald's grave and lay down. Maybe when Mrs. Bald came down from the clothesline, she would tell him what to do.

He sniffed. Even through a stuffed nose, he thought he smelled garbage. Or was it eggs?

Meanwhile, Dog Alfred's sneeze flew around the farm, then into the sky.

The man in the moon was just beginning to yellow up for the night. The sneeze snuggled against the man and whispered in his ear. The man in the moon frowned, inspecting the trodden field. Fixing this would take time.

Below, on the farm, in front of the clothesline, Bones knelt before his mother, his hands outstretched, his head bowed. Fat pumped his arms as though he had conquered the world. And the cat turned toward Fat.

I've always loved the texture of
fairies. They crunch so wonderfully.
Better than mice. And their wings,
so airy and light, taste a bit like
cotton candy. You may not have
known that cats like cotton candy,
but we do. we like it almost as
much as we like fairies.

After swallowing Fat, I smiled
innocently at Mrs. Bald and backed
up against Bones's face. I stuffed
what was left of my tail up his nose
and closed his mouth with my paws.
He was too beat down at that
point to put up much of a fuss. He
jerked once or twice, then lay still.

All was quiet.

With the sounds of war gone, Mrs. Bald's tear ducts dammed up. Little by little, water filled her skin, and it didn't take more than two hours for her to be just as she had been. I licked her hands. She always liked it when I did that.

The first thing Mrs. Bald did when she unpinned herself from the clotheslines was to fill my shiny red food dish with a fresh batch of pig foot stew.

Like most cats, I'm a philosopher. Like most philosophers, I try to make sense of chaos. Some moments of this tale are conjecture, I admit, but of this I am certain— with Mrs. Bald as flat as one of those pancakes Bones loved so well, my shiny red food dish would have remained empty. And I know from experience, an empty food dish means chaos. By getting rid of Fat and Bones, I restored order. I restored peace. Not bad for a tailless cat, if I do say so myself.

Author Acknowledgments

This is the part of the book I was most looking forward to writing, the part where I can thank the many people who have been patient and kind during this story's journey to publication.

Mom and Dad, thank you for being exemplary parents and teachers, for living with conviction, and teaching us kids that love will make a difference. My siblings, Shaelyn, Shannon, Luke, and Josh—thank you for all the Dutch rubs and ribbing. I owe you one. Jane, Uma, Shelley, and Rita—you are gifted teachers with wide-open hearts; thank you for freely sharing your wisdom and taking the time. My intrepid agent, Linda Pratt—you took a chance— thank you, thank you for knowing where this book needed to be. My thanks also to my super-smart editor Greg Hunter, who guided the many strands of this story into a coherent whole. And to the entire team at Lerner, most especially to Andrew Karre, who welcomes the unusual. To the Super Secret Society of Quirk and Quill; my peeps, you are true friends. Adam Doyle, you found just the right tone; thank you. Edit Ford, I'm so grateful for your humor and encouragement. Bear and Boo—my peanuts, I love you. And Philip, you know all of it—thank you.

About the Author

Having recently returned to the United States after living six years in China, Larissa Theule resides in Pasadena, CA, with her family and dog. Once a barista, a billing assistant, and a high school teacher, she now devotes her time to writing stories and growing a rambling garden, where no fairies have been sighted— at least not yet. *Fat & Bones* is her first book.

About the Illustrator

Adam S. Doyle grew up in Boston and later earned his BFA from The Rhode Island School of Design and his MFA from The School of Visual Arts in New York City. So far he's also lived in Los Angeles, Rome, Auckland, and Hong Kong.

Adam draws inspiration from nature, mythology, the human condition, and the magic of creation—the way a splatter of paint transforms the blank page into a living being or vast landscape. Between these two worlds, the real and the imagined, he keeps the window partly open. In addition to picture books, Adam's work appears in galleries, on cards, on bodies, on trucks, and on book covers, including Maggie Stiefvater's bestselling Raven Cycle. You can see more at adamsdoyle.com.